MILESTONES
OF SPACE

ELEVEN ICONIC OBJECTS
FROM THE SMITHSONIAN
NATIONAL AIR AND SPACE MUSEUM

MICHAEL J. NEUFELD

WITH CURATORS FROM THE
NATIONAL AIR AND SPACE MUSEUM

Smithsonian
National Air and Space Museum

Washington, D.C.

In association with

First published in 2014 by Zenith Press, a member of Quayside Publishing Group, 400 First Avenue North, Suite 400, Minneapolis, MN 55401 USA

Zenith Press titles are also available at discounts in bulk quantity for industrial or sales-promotional use. For details write to Special Sales Manager at Quayside Publishing Group, 400 First Avenue North, Suite 400, Minneapolis, MN 55401 USA.

To find out more about our books, visit us online at www.zenithpress.com

ISBN-13: 978-0-7603-4444-6

Library of Congress Cataloging-in-Publication Data

National Air and Space Museum.
  Milestones of space : eleven iconic objects from the Smithsonian National Air and Space Museum / Michael J. Neufeld, editor.
      pages cm
  Includes bibliographical references.
  ISBN 978-0-7603-4444-6 (hardback)
 1. Astronautics--United States--Equipment and supplies--Pictorial works. 2. National Air and Space Museum--Catalogs. I. Neufeld, Michael J., 1951- editor of compilation. II. Title.
  TL506.U6W376 2014
  629.4074'753--dc23

Editor: Elizabeth Demers
Design Manager: James Kegley
Layout: Diana Boger
Cover Designer: Jason Gabbert
Design: Karl Laun

On the front cover: Neil Armstrong's A7-7 spacesuit. *NASA*

Printed in China

10 9 8 7 6 5 4 3 2 1

CONTENTS

 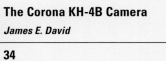

ABOVE: ABOVE: The National Air and Space Museum on the National Mall at dawn. *NASM*

RIGHT: The Museum's Steven F. Udvar-Hazy Center at dusk. *NASM*

FOREWORD AND ACKNOWLEDGMENTS

THIS BOOK PRESENTS a capsule history of the U.S. space program through eleven key artifacts held by the Smithsonian's National Air and Space Museum (NASM). It does not pretend to be comprehensive, but we, the chapter authors, feel that these iconic objects represent the major accomplishments and programs of the United States in space. With one exception, these essays are by the curators in the Space History Division, who currently have the artifacts in their collections. The essay on Skylab was contributed by Stuart W. "Bill" Leslie of Johns Hopkins University, who was the museum's Charles A. Lindbergh Chair in Aerospace History in 2012–13, and Layne Karafantis, one of his graduate students. In addition, one sidebar to the *Friendship 7* essay was adapted from a museum web publication by Teasel Muir-Harmony, a graduate student at MIT and a former fellow and employee, who kindly gave her permission.

The authors would particularly like to thank Edgar Durbin, a volunteer in the Space History Division, who took on the major job of assembling the photographs for the chapters. He did extensive picture research, and he worked to secure higher-resolution images when possible. He also took the responsibility of uploading the final versions of the text and picture files to the publisher's website. The amount of work he contributed to the final product was far more than what one would expect of a volunteer. He has made an outstanding contribution.

We also would like to thank Trish Graboske, the museum's publications officer, and Paul Ceruzzi, the chair of the Space History Division, for initiating this project. Moreover, the NASM Archives Division, headed by Marilyn Graskowiak, had a central role in supplying many of the photographs in the book. The chief photo archivist, Melissa Keiser, was particularly helpful. Two of NASM's photographers, Dane Penland and Eric Long, shot new images of artifacts for this book, and they and Mark Avino also took virtually all of the older object portraits. Stamatios M. Krimigis of the Johns Hopkins University Applied Physics Laboratory provided helpful advice on the Voyager Interstellar Mission. Finally, Jane Odom, chief archivist in the NASA History Office, and other NASA employees, were very helpful to Edgar Durbin and the authors in their research.

—Washington, D.C.
January 2014

ON FEBRUARY 20, 1962, astronaut John H. Glenn Jr., became the first American to orbit Earth. He made three trips around the world in *Friendship 7*, a small spacecraft weighing barely more than a ton and a half, which had been hurled into orbit by an Atlas intercontinental ballistic missile (ICBM). Glenn was the fifth person in space, following two Soviet cosmonauts who had orbited and two American astronauts who had made suborbital trips of fifteen minutes. Because he was the first to match the Soviet achievement of orbiting, his fame in the United States quickly eclipsed even that of the first American in space, Alan B. Shepard. As a result, *Friendship 7* was chosen for the Milestones of Flight gallery at the National Air and Space Museum when it was opened in 1976—not Shepard's *Freedom 7*.

Project Mercury, the name given to the initiative to put an American in orbit, had its origin in the immediate aftermath of the Soviet *Sputnik* surprise of October 4, 1957. The orbiting of a satellite immediately legitimized the Soviet claim of having tested an ICBM—thus making the nuclear threat much more real. *Sputnik* also hurt American pride as it seemed to demonstrate Soviet scientific and technical superiority, provoking an outcry in the press and in Congress against President Dwight Eisenhower's administration. When the Soviets sent a dog into orbit in early November, and the first U.S. satellite launching attempt ended in an embarrassing launch pad explosion in early December, it only strengthened the public and scientific outcry. Eisenhower reluctantly conceded to the demands for an accelerated national space program.

MERCURY CAPSULE
FRIENDSHIP 7
1

This green light only exacerbated the rivalries among the military services, who vied for a place in space, as well as in long-range missile programs. The army and air force started competing plans for launching humans. Annoyed, and advised both by politicians and scientists that a civilian space agency would best represent America's international image in the Cold War, Eisenhower proposed the National Aeronautics and Space Administration (NASA), which was to be built on the foundation of the National Advisory Committee for Aeronautics (NACA). The president also made the new agency responsible for the first experimental human space program. When NASA came into existence on October 1, 1958, it created Project Mercury to put one man into orbit. At the time, only men were considered for the job that NASA soon dubbed "astronaut"; women pilots who later privately took the astronaut medical exam were rebuffed by the agency.

Pre-*Sputnik* space advocates imagined that humans would travel to orbit in multi-person spaceplanes. But in a mad dash to beat the Soviets, developing a spaceplane

Friendship 7 in the Milestones of Flight gallery at the National Air and Space Museum. *NASM*

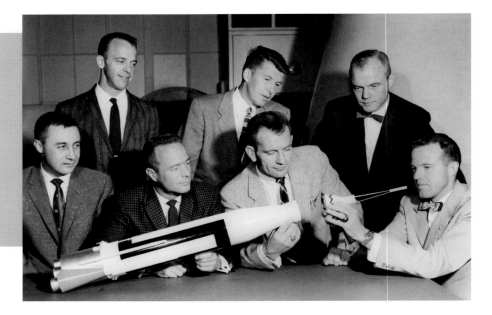

An early publicity shot of the seven Mercury astronauts, probably taken soon after their selection was announced in April 1959. From left to right, sitting: Virgil I. "Gus" Grissom, M. Scott Carpenter, Walter "Deke" Slayton, and L. Gordon Cooper Jr. Standing: Alan B. Shepard Jr., Walter M. Schirra, and John H. Glenn Jr. *NASA/NASM*

was a luxury the United States could not afford—at least for the first program. The quick-and-dirty approach was to create a ballistic capsule, essentially a missile nosecone big enough to hold a passenger and the systems needed to keep him alive and get him back to Earth. At the NASA Langley Research Center in Hampton, Virginia, aeronautical engineer Max Faget had already drawn a conical spacecraft with a cylindrical nose. The broad, flat base was covered by a heat shield that would create a shock wave to slow the spacecraft down and ward off the intense heat that would be created by plowing into the Earth's atmosphere at only slightly less than the orbital speed of 17,500 mph. Protecting the capsule was technology directly derived from ICBM warheads, either heavy metal "heat sinks" or fiberglass and resin "ablative heat shields" that would erode during the reentry. The latter type was used on *Friendship 7* and all Mercury orbital missions.

The only big rocket the United States had in 1958 that could launch such a spacecraft was the air force's Atlas, the nation's first ICBM. But the Atlas was still early in its development and had an alarming tendency to blow up. Robert Gilruth's Space Task Group, which was based at the Langley Center and was responsible for Project Mercury, therefore decided that early space missions to test the capsule and train the astronauts were advisable. Adopting an idea from the army's short-lived human space project, the task group also decided to use that service's Redstone tactical ballistic missile, which could send the capsule up to a hundred miles in altitude but could achieve only about one-third the velocity needed to reach orbit.

In April 1959, NASA introduced its first seven astronauts who, by presidential order, were chosen from the pool of active military test pilots. Not wanting to be thought of merely as passengers, the astronauts moved quickly to put their stamp on the program. The Mercury capsule was being designed for automatic operation, in large part because the medical effects of spaceflight were unknown. Some doctors imagined that weightlessness and isolation would create physical or psychological

JOHN HERSCHEL GLENN JR.

PILOT AND POLITICIAN

Before John Glenn was a famous astronaut, he was a well-known marine test pilot. In Project Bullet, he set a transcontinental speed record in July 1957 of 3 hours and 23 minutes, flying a F8U Crusader. It was the first coast-to-coast supersonic flight, an accomplishment that probably got him his national appearance on the TV game show *Name That Tune* in the fall of 1957.

Born on July 18, 1921, in Cambridge, Ohio, Glenn grew up and went to school in nearby New Concord. In 1942 he volunteered for the Marine Corps, and before he went overseas he married his childhood sweetheart, Anna Castor. They later had two children. Glenn flew fifty-nine combat missions in an F4U Corsair in the Pacific in 1944–45. During the Korean War, he served two combat tours in jet fighters, first in F9F Panthers in the marines and then in the F-86 Sabrejet on an exchange tour with the air force. Near the end of the war, he shot down three MiG-15s. In 1956, he was assigned to the Naval Test Pilot School at Patuxent River, Maryland, and was serving as a test pilot for the Navy Bureau of Aeronautics when he was chosen as one of NASA's seven Mercury astronauts in 1959.

The heroic status he acquired from his flight in *Friendship 7* brought him close to President John F. Kennedy and the Kennedy family and made it even more unlikely that he would ever be assigned to another space mission. Glenn resigned from the astronaut corps in January 1964, as he had decided to run for office as a Democrat. He retired from the marines as a colonel a year later, shortly after losing an election to become U.S. senator from Ohio. After a stint as a corporate executive, he finally was elected to the Senate in 1974, and he served until the beginning of 1999.

At the end of his final term, Glenn got a chance to return to space when his campaign for a space shuttle mission bore fruit. He was assigned to STS-95, with the purpose of researching the effects of spaceflight on the aging body (he was seventy-seven). He flew nine days, from October 29 to November 7, 1998, in the shuttle *Discovery* as a payload specialist. He still holds the record as the oldest person ever to fly in space. For his accomplishments in space and on the ground, he received a Congressional Gold Medal in 2009 and a Presidential Medal of Freedom in 2012.

Marine Maj. John Glenn with the Vought F8U Crusader aircraft he used in Project Bullet to set a transcontinental speed record in July 1957. *NASM*

problems that would incapacitate the astronaut. Unimpressed by these scenarios of doom, the seven intervened to have a window placed above the pilot's head and requested various changes to the cockpit layout that would increase control if the automatic systems malfunctioned.

Although the Mercury program proved longer and more complicated than expected, by the end of 1960 the Space Task Group launched the first unmanned Mercury-Redstone suborbital mission, followed by chimpanzee Ham in January 1961

The headline in *The Huntsville* (Alabama) *Times* after the successful one-orbit flight of Yuri Gagarin on April 12, 1961, shows the impact of the first human spaceflight on world opinion. *NASM*

The Huntsville Times

Man Enters Space

'So Close, Yet So Far,' Sighs Cape

Soviet Officer Orbits Globe In 5-Ton Ship

(the doctors required one such live-subject mission before the first human suborbital and orbital flights). While Ham returned safely, any hope that the United States would be first in space was dashed on April 12, when Yuri Gagarin made one orbit around Earth in the Soviet spacecraft *Vostok 1*. On May 5, Alan Shepard accomplished his suborbital flight, followed in July by Virgil "Gus" Grissom. John Glenn acted as backup astronaut for both. Gherman Titov spent an entire day in orbit in early August. But even without the new Soviet triumph, it was clear to Gilruth's team that another suborbital

test would just delay the orbital flights without producing much new knowledge. It was time to move on—giving Glenn the chance to become the first American in orbit.

In September 1961, a capsule made one orbit on the Mercury-Atlas 4 mission, with a "crewman simulator" inside to test the environmental control system. The capsule had been refurbished after its escape system had rescued it from an exploding booster in April, an indication of how dangerous such a flight was for the astronaut. The next mission featured the chimp Enos, who went around the Earth twice in November before the attitude control system used up too much hydrogen peroxide fuel for the control jets, forcing a mission termination before the nominal three-orbit Mercury mission could be completed. Enos's return to Earth was enough to finally give John Glenn the go-ahead to fly on Mercury-Atlas 6.

Glenn decided to name his spacecraft *Friendship 7*, following the tradition established by Shepard, who had added the number seven to signify the members of the astronaut team. But technical problems soon pushed the flight into January 1962, which was only the beginning of the mission's problems. The patience of the press and the public soon wore thin after several more postponements, many due to cloudy winter weather at the launch site: Cape Canaveral, Florida. After the unmanned Mercury-Atlas 1 had exploded in 1959, NASA had instituted a rule that launches must be visible for filming, so that accidents might be more easily investigated.

Finally, on February 20, the weather began to improve. Glenn rose at 2:20 a.m. Eastern time, had breakfast, suited up, and shortly after 6:00 entered the spacecraft. (So tight was the fit that the astronauts joked: "You don't get into it, you put it on.") Technicians strapped him into his form-fitting fiberglass couch designed to make the high "G forces" of acceleration and deceleration more tolerable. Several more delays due to mechanical problems ensued, but at 9:47 the Atlas rumbled to life.

Enduring an acceleration of up to eight Gs (eight times Earth's gravity) pressing on his chest, in five minutes Glenn was in an orbit of 162 by 100 miles, traveling

LEFT: Glenn enters *Friendship 7* on the morning of the launch, February 20, 1962. Getting into the cramped cockpit was a very tight fit and required the assistance of a couple of technicians. *NASA/NASM*

RIGHT: The launch of *Friendship 7* on the Mercury-Atlas 6 mission. The booster was a modified ICBM. On the top of the spacecraft is the "escape tower." It has a powerful solid-fuel rocket to yank the capsule away in the case of an emergency. As both the Mercury-Atlas 1 and 3 boosters had exploded in unmanned tests, a successful launch was far from assured. *NASA*

17,544 mph. Immediately after it separated from the booster, *Friendship 7*'s automated systems turned it around so that the retrorocket pack strapped to the heat shield pointed in the direction of flight, enabling an emergency return at almost any time. Glenn thus rode backward around the world. Out the window, he could see the Atlas slowly tumbling in orbit behind him. Within twenty minutes, he crossed the coast of West Africa and talked to the tracking stations in the Canary Islands and Kano, Nigeria. These stations were linked together in a global network run by Mercury Control, which was situated at the Cape.

Over the Indian Ocean, about halfway through his eighty-nine-minute orbit, Glenn experienced his first of three orbital sunsets. He described it afterward as:

> a beautiful display of vivid colors. . . . As the sun gets lower and lower, a black shadow moves across the earth until the entire surface that you can see is dark except for the bright band of light along the horizon. At the beginning, this band is almost white in color. But as the sun sinks deeper the bottom layer of light turns to bright orange. The next layers are red, then purple, then light blue, then darker blue and finally the blackness of space. They are all brilliant colors, more brilliant than in a rainbow. . . .

When he crossed the Australian coast a little while later, he told the "capcom" (capsule communicator) at Muchea, astronaut Gordon Cooper, that he could see lights. The citizens of the cities of Perth and Fremantle had turned everything on to greet him.

Twenty minutes later, out in the middle of the Pacific, riding backward, he saw a sunrise via the periscope screen in the middle of his control panel. The periscope,

BELOW: A photo taken by an on-board camera shows Glenn during the flight. On his chest was a mirror in which the control panel was reflected, if Mercury engineers later wanted to relate the photograph to onboard events. Another movie camera over his right shoulder took photos of the instrument panel. *NASA/NASM*

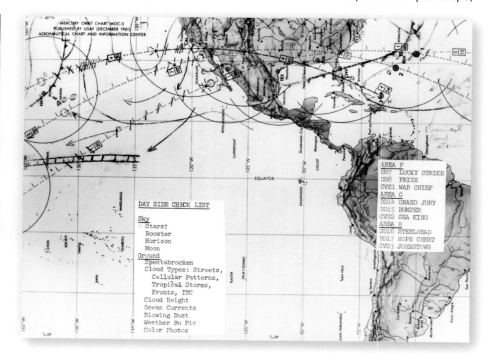

RIGHT: Part of the world map that Glenn carried with him during his mission, showing orbital tracks and the radio reception range of tracking stations. *NASM*

which had a wide-angle lens, could be extended out of the bottom of the spacecraft as a covering door moved out of the way. It had been the primary visual display built into the spacecraft before the astronauts pushed for a large window. Just as the sun came up, Glenn witnessed a "startling sight" when he glanced up at the window. "All around me, as far as I could see, were thousands and thousands of small, luminous particles." He called them "fireflies." During and after the flight, he considered it the greatest mystery of his spaceflight experience. It was only when his backup, Malcolm Scott Carpenter, flew the same three-orbit mission in May that the phenomenon was resolved. Carpenter banged the wall of the spacecraft and produced more. They were tiny ice particles that had been created by the evaporator in the environmental control system and had adhered to the outside of the capsule.

As Glenn approached the Mexican coast near the end of the first orbit, he began to experience his first attitude control problems. The nose drifted to the right. When it got to about 20° off the center line, a thruster jet in the cylindrical nose section fired to push it back, and then the opposite thruster fired to stop the motion. The drift, caused by a fuel leak, would then begin all over again. This behavior began to eat up the supply of hydrogen peroxide in the automatic system. Soon thereafter, one of *Friendship 7*'s thrusters failed intermittently.

Glenn experimented with his options. He had a separate manual control system with another set of tanks and jets. He also had a mode called "fly-by-wire," whereby

The cockpit of *Friendship 7*. The astronaut's attitude control stick is at right, and on the left is the "abort handle" for triggering an emergency escape from the launch vehicle. Visible in the lower center of the main control panel is the screen for the periscope, and to the left is a row of indicator lights for various important functions. The last one is the light for the landing bag deployment, which became a critical issue during the mission. *NASM*

THE ORIGINS OF MISSION CONTROL

The Mercury Control center at Cape Canaveral, Florida, is shown during Glenn's mission. The three orbits are inscribed on the map, as are the range of the various tracking stations for Glenn in his orbit not much more than 100 miles (160 km) above the Earth. *NASA*

Christopher Columbus Kraft Jr., a young NASA engineer from tidewater Virginia, was the visionary of the mission control concept—then called Mercury Control. Early on, in thinking about how they were going to communicate with an astronaut circling the Earth, it became clear to Kraft that NASA needed a worldwide network of tracking stations linked to a central control center. Such a center would allow for many functions of the capsule to be backed up from the ground and for commands and emergency procedures to be communicated to the pilot. Kraft decided that he would differentiate the various functions of the spacecraft, booster and network into desks, which originated many of the roles and acronyms of mission control.

he manually controlled the automatic system through the stick in his right hand. That seemed to work the best. Controlling his own attitude allowed him to sometimes stop flying backward, and it demonstrated the value of a "man in the loop." Enos's flight had been terminated because of a similar failure, but with Glenn taking over, he felt it should be possible to complete the three-orbit mission. The one thing he had to watch was fuel quantity, as he needed to be able to control the capsule during the reentry phase.

While going around the Earth, Glenn carried out various experiments. Periodically, he had to inflate the blood pressure cuff that was built into his suit.

For example, the controller responsible for the launch vehicle was called "Booster" and had his own console for that function; the person responsible for communication with the astronaut (always himself an astronaut) was called "CAPCOM" (Capsule Communicator).

Creating Mercury Control between 1959 and 1961 presented many challenges. There were no geostationary communications satellites yet. Undersea telephone cables and teletype machines had to be used to link up the network. Computers existed only as huge mainframes that took up entire rooms. Orbital and trajectory calculations were made on an IBM machine at the NASA Goddard Space Flight Center, outside Washington, D.C., which also served as the center of the communications infrastructure; these calculations were then communicated to Mercury Control at the Cape. Tracking stations sometimes had to be built in remote and often dangerous regions, requiring the assistance of the State and Defense Departments. The Kano (Nigeria) and Zanzibar stations (the latter on a small island off the East African coast) proved especially troublesome due to anti-American feelings locally. Both stations were removed after Mercury. Moreover, Kraft's deputy, Eugene Kranz, had to train the key members of all tracking stations in how to function as one team.

The unmanned orbital missions in fall 1961 gave Mercury Control its first global tests. But it was the crisis over the false heat shield signal on Glenn's flight that showed Kraft that he had to establish the authority of the Flight Director—during Mercury, almost always him—to make decisions without external interference. It also demonstrated the critical importance of clearly defined mission rules. In hindsight, Kraft decided that he should never have allowed the untested and perhaps dangerous procedure of keeping the retrorocket package on during the reentry.

Mercury Control became Mission Control in Houston when Project Gemini began, following the transfer of the human spaceflight program to Texas. The last mission controlled out of the old building at the Cape was the three-orbit flight of *Gemini 3* in 1965, the first to carry two astronauts.

The tracking icon used on the world map in Mercury Control is now an artifact in the collection of the National Air and Space Museum. *NASM*

He also practiced with an eye chart, and he tried the squeeze-tube apple sauce. He had two cameras, one for taking daytime pictures of the Earth and a special camera with a quartz lens for taking ultraviolet photographs of the constellation Orion during the night, an observation not possible from Earth because of the screening effect of the atmosphere. In order to make the camera usable with bulky spacesuit gloves, it was outfitted with a pistol grip. But after the attitude control problems began on the second orbit, he had to drop many additional planned astronomical and Earth observations.

FRIENDSHIP 7's "FOURTH ORBIT"

Three months after its flight, *Friendship 7* began its second mission, or what was popularly referred to as its "fourth orbit." A worldwide exhibition in nearly thirty cities was organized to promote the United States and its space program. Ever since *Sputnik*, the United States and USSR had come to see spaceflight as a leading means for demonstrating power, technological capability, and national values. As a result, the U.S. space program, and its exhibition abroad, became important instruments in American foreign relations during the Cold War.

Over the course of its three-month-long tour, *Friendship 7* was seen by roughly four million people. Another twenty million people watched television programs about the capsule broadcast from the exhibition sites. In early May 1962, on the first day that the capsule was displayed at the Science Museum in London, thousands of people had to be turned away because the huge crowds overtaxed the facilities. Even though tropical thunderstorms drenched Nigeria and an earthquake shook Mexico during the capsule's visit, the exhibit caused a much larger stir in every city it visited than officials at NASA and the State Department had ever imagined.

The capsule was flown around the world in a U.S. Air Force cargo plane emblazoned with the words "around the world with *Friendship 7*" and a map of the four continents that the capsule visited over the summer. A member of NASA's Cape Canaveral staff accompanied the craft to answer questions.

Although *Friendship 7* drew record crowds in Europe and Africa, the capsule received its most overwhelming response in Asia. When it arrived in Bombay in mid-July, fifty thousand residents waited for up to four hours to see the display at a stadium. In downtown Tokyo, the

Unbeknownst to Glenn, back in Mercury Control there was a growing sense of alarm, or at least discomfort, at a telemetry signal coming from the spacecraft. Called a "Segment 51," it was an indication that the heat shield could possibly be loose. The Mercury capsule's shield was attached to a "landing bag" that dropped down during the last part of the parachute descent before ocean splashdown. That landing bag provided a cushion against impact and acted as a sea anchor to increase the capsule's stability in the ocean. Three latches held the heat shield to the main body of the spacecraft. If the heat shield really was loose as a result of the latches releasing, the only thing holding it on were the three straps that held the retrorocket package in place against the shield. These were attached to the rim of the main body, which itself was covered with special, high-temperature metal shingles made of a nickel alloy or beryllium. Flight Director Christopher Kraft felt immediately that the loose heat shield signal had to be false. Checking spacecraft schematics in Mission

capsule was taken to Takashimaya, a leading department store. Several hundred police and guides directed the crowd into a line that climbed nine flights of stairs, zigzagged across the roof of the building, and then descended back down nine flights of stairs to the first floor where the capsule was displayed. Over the course of its four-day visit, more than five hundred thousand people came to see *Friendship 7*.

A year after his flight, John Glenn wrote McGeorge Bundy, President Kennedy's national security advisor, that the tour showed that the American space program "was not just a propaganda effort . . . , but a well-thought-out scientific program that could eventually benefit all peoples of the world as the scientific exploration it is." The openness of the exhibit stood in for the nation and its political ideology. When *Friendship 7* was laid bare before people from around the world, it strengthened the impression that the U.S. program was real, benign, apolitical, and designed for the benefit of all mankind.

—*Teasel Muir-Harmony*

Control, it was apparent that the way the landing bag was wired, one errant latch sensor would be enough to create the signal. But there was no way to prove one way or the other whether the problem was real.

Glenn first became aware that there was a concern when he passed over Australia for the second time. Cooper asked him if the landing bag switch was off and whether he heard any "banging noises" when the attitude changed at "high rates." "Negative" was Glenn's answer. Then he pretty much forgot about it, as the ground did not raise it again for more than an entire orbit. Much of his time was spent dealing with attitude control problems, errors in his gyroscopes, and carrying out required medical tests. But tension built at Mercury Control as discussions continued about what to do. Kraft's inclination was to go for a normal reentry, which meant jettisoning the retrorocket package after the three solid-fuel rockets fired to take Glenn out of orbit. Capsule designer Max Faget and Mercury operations director Walter Williams,

The specially modified ultraviolet camera used by Glenn to take astronomical photographs of the constellation of Orion during the first orbit. It is currently on exhibit at the National Air and Space Museum, as is the spacecraft. *NASM*

A close-up of the landing bag light on the control panel shown on page 13. *NASM*

Kraft's boss, argued they would kill the astronaut if Kraft was wrong and the heat shield really was loose. Kraft replied that if all the solid fuel in the retrorockets did not burn, the package could blow up during the reentry. In any case, the aerodynamics of leaving it on were unknown.

That discussion went on almost until the moment of retrofire, which was to occur over California near the end of the third orbit—the designated landing zone was in the Atlantic east of the Turks and Caicos Islands. When Glenn passed near the Hawaii ground station immediately before that critical event, the capcom conveyed Mercury Control's request to turn the landing bag switch to "Auto" and see if he got a light, the very last one on the row of indicator lights on left side of his instrument panel. Glenn did that with a little trepidation, wondering if it might accidentally deploy the landing bag, but the light did not come on and he quickly turned the switch off. The capcom told him that the reentry sequence would be normal—but that prediction proved premature.

When Glenn came in range of the California tracking station at Point Arguello, astronaut Wally Schirra talked him through the retrofire sequence and told him to leave the pack on until he reached the Corpus Christi, Texas, station. Before that command, a heated discussion had occurred in Mercury Control. Kraft had finally given in to the pressure from Williams and Faget. When Glenn, already descending, passed over Texas, he was told to override the sequence and keep the package on all the way through the reentry. He had had three good retrorocket burns, meaning, presumably, that all propellant in them had burned and the package would not explode. Now Glenn was really concerned—as was the outside world, which had been alerted to the possibility of a loose heat shield by air force Lt. Col. John "Shorty" Powers, the public information officer for Project Mercury and the voice of Mercury Control to the networks. Would America's first orbital mission end with the heroic astronaut consumed in a ball of fire?

Glenn himself stated in a no-doubt ghostwritten account in the book published by the astronauts that same year, 1962, "I knew that if the shield was falling apart, I would feel the heat pulse first at my back, and I waited for it." Red-hot glowing chunks of the retro pack went by his window and one of the straps broke and flapped in front of him. Now cut off from radio communications by the ionization of the air around the spacecraft—the first communications "blackout" ever experienced by an American astronaut—he could only talk to the on-board recorder. When it was at its worst, he commented: "A real fireball outside." The seconds seemed to drag on like hours, as they did in Mercury Control. The blackout lasted nearly four and a half minutes. The heat pulse at his back never came. Eventually, he slowed down enough that the blackout ended and he could again talk to Al Shepard, capcom in the Cape control center. Everyone heaved a sigh of relief.

As he descended, the heat soaking in from reentry made the cabin too warm, and the capsule began oscillating wildly in the lower atmosphere. Glenn used up the fuel he had left in his two control systems to try to manage it. He considered firing the mortar that sent out the small drogue parachute that was supposed to stabilize the spacecraft, but as he was reaching for the switch, it deployed automatically. At

Sailors on the destroyer USS *Noa* prepare to haul *Friendship 7* aboard ship. *NASA/NASM*

10,800 feet the main parachute unfurled, and at 4 hours, 55 minutes, and 23 seconds, mission elapsed time, he hit the water. It was about 2:42 p.m. Eastern time. Because of the Mercury program's inexperience with reentry, he landed about forty miles short of the aircraft carrier, but quite near the destroyer *Noa*. While Glenn sweated profusely in the stifling cockpit, the destroyer pulled alongside and hoisted *Friendship 7* onto the deck. Finally onboard, Glenn fired the explosive hatch to make a quick exit. He was drenched and dehydrated but elated—mission accomplished.

The successful flight provoked a national reaction of joy and pride that surprised even Glenn and the members of Project Mercury. President John F. Kennedy called the ship to congratulate him and traveled to Cape Canaveral for the ceremonies. On the 25th, a quarter-million people attended a parade in Washington, D.C., and the president addressed a joint session of Congress. March 1 was John Glenn Day in New York City, with a huge ticker-tape parade. As for the spacecraft, after a thorough technical evaluation, it made a tour around the world in 1962, reaching the Smithsonian in November. It quickly overshadowed *Freedom 7*. Probably because Glenn's mission was the achievement that at least equaled the Gagarin flight in the American mind, *Friendship 7* became *the* symbol of the Mercury program and of the entry of American astronauts into space.

—*Michael J. Neufeld*

FRIENDSHIP 7
SPECIFICATIONS

MANUFACTURER:
McDonnell Aircraft, St. Louis
LENGTH (AT NASM):
9 ft. 4 in. (2.84m)
LENGTH (WITH ESCAPE TOWER):
26 ft. (7.90m)
LENGTH (IN ORBIT):
11 ft. 4 in. (3.45m)
DIAMETER (AT BASE):
6 ft. 2 in. (1.88m)
WEIGHT (CURRENT):
2,987 lbs. (1,358 kg)

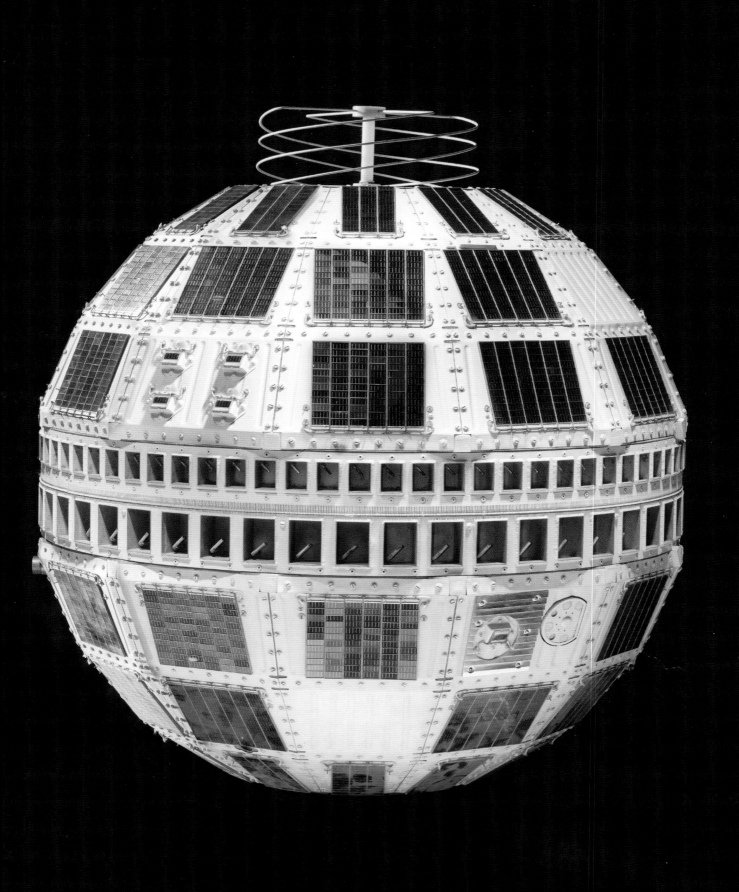

I **N THE EARLY YEARS** of spaceflight, nearly each accomplishment of the U.S. civilian space program—whether of a human spaceflight or a satellite or probe placed beyond the atmosphere—drew wide attention. Each new milestone heralded scientific or technical advances and stood as a symbol of the dramatic changes occurring in postwar America, in the harrowing geopolitics of the Cold War, or in perceptions of possible futures for human society. Spaceflight at once burnished longstanding narratives of progress and prompted concerns that such advanced capabilities were leaping ahead of society's readiness to adapt or benefit from such changes.

The communications satellite Telstar, launched on July 10, 1962, was no exception, generating intense interest, excitement, and commentary for the next several months before its orbital path and electronic problems diminished its effectiveness. Its name—a contraction of the words "telecommunications" and "star"—perfectly captured the fascination of the moment. As a creation of the American Telephone & Telegraph Company (AT&T), then the largest business in the world, the satellite offered a contrast to the space projects sponsored by NASA and the U.S. defense establishment. Through summer and fall, the small spacecraft—a sphere less than three feet in diameter—had an outsized impact on politics and on the public imagination in the United States and Europe. As a new, space-age tool of communications, it greatly enhanced the capability to link the continents, promising new possibilities for human

The Museum's artifact is a backup spacecraft for Telstar 1 and Telstar 2 (launched on May 7, 1963). It arrived at the Smithsonian in its original container, specially designed for shipment from Bell Telephone Laboratories to Cape Canaveral and also for testing the satellite at the launch site. But the satellite never took this journey as both Telstar 1 and 2 reached orbit. *NASM*

TELSTAR /2

In summer 1962, Telstar shared public attention with Project Mercury human spaceflights and with those of the Soviets. This postal cover links the two American efforts, suggesting the broad public interest in space accomplishments. Indeed, "space mania" gripped the stamp world in 1962; in November, twenty thousand collectors packed New York City's Fourteenth National Postage Stamp Show to view *Friendship 7* and a full-sized model of Telstar. *Smithsonian National Postal Museum*

connection and understanding. Yet, Telstar provoked anxiety about whether increased flows of information, particularly via television, helped or hindered a world already made smaller, more interdependent, and more fragile through the Cold War's threat of nuclear annihilation. The stakes associated with communications, in its helpful or negative potentialities, never seemed more profound.

In its blaze across these months, Telstar helped shape perceptions of a variety of concurrent events: U.S. and USSR human space flight missions, a string of nuclear weapons tests, global economic problems, and, on the other end of the spectrum, the role of American popular culture, here and abroad, especially as presented on television. In weaving these multiple historical threads, the satellite became a technical, political, and cultural happening, ushering in the world of globalized information we take for granted today.

After launch, Telstar assumed an elliptical orbit positioned to enable communication between North America and Western Europe. As an "active" satellite, its distinctive technical feature was its ability to receive a radio signal from a ground station and then immediately retransmit it to another, hundreds or thousands of miles away. This capability had been accomplished in rudimentary fashion with the Courier communications satellite, launched in 1960. Telstar, though, was more robust, possessing sufficient power and radio signal bandwidth to transmit across the Atlantic not only phone calls and faxes but high-speed data and, most notably, television, which only in recent years had become a fixture in everyday life. A key objective of the satellite project had been to demonstrate this latter possibility—a communications first.

Telstar was experimental—as were the few other communications satellites already launched or in the planning stages. Scientists did not fully know either the space environment or the best methods for passing radio waves to space and back. They needed to perfect satellite equipment that would operate in the harsh conditions of space, as well as ground-based methods for tracking and control. As one news account noted, one daunting task was to "pinpoint an object 34 inches in diameter moving faster than a bullet more than three thousand miles out in space." The "bullet" metaphor understated the challenge: Telstar traveled at more 16,000 mph. Not least, the Van Allen Belts, discovered in 1958 by Explorers I and III and which were filled with high-energy particles, seemed a potential hazard for satellites (and for future human spaceflight missions). Telstar tested this concern, with its orbit designed to pass regularly through the Belts.

Telstar was composed of several key systems. The primary communications devices were the equatorial antennas (the double band of recesses girdling the satellite) and the traveling wave tube. These antennas received signals from a ground station, sent them to the traveling wave tube—which amplified the signals ten billion times—and then passed them back to the antennas for retransmission to the ground. The antenna at the top received operational commands and sent information (telemetry) on the status of the satellite's operations. The solar cells, working in combination with the battery, provided the satellite's power. *AT&T Archives and History Center*

ANTENNA (COMMAND AND TELEMETRY)

TELEMETRY MODULE

SOLAR CELLS

TRAVELING WAVE TUBE AMPLIFIER

EQUATORIAL ANTENNAS

NICKEL-CADMIUM BATTERY

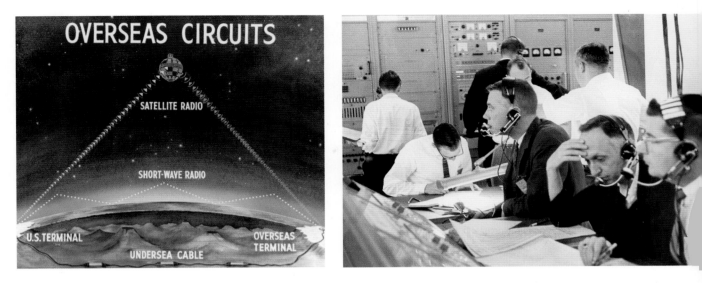

Telstar, of course, was not just the spacecraft. It stood as a testament to large-scale systems engineering and the coordinated work of teams of experts in the U.S. and Europe. In addition to the satellite itself, the system included massive and complex ground stations: in Pleumeur-Bodou, France; at Goonhilly Downs, England; and, on the U.S. side, at a primary ground station in Andover, Maine, built by AT&T, which also assisted the French station. While Telstar orbited, these ground stations collectively provided the means for transatlantic satellite communications, creating a greatly expanded conduit between the two continents. A few years before Telstar, in 1956, AT&T inaugurated a new undersea cable, linking the United States and Canada with England, for telephony and data—but even so the technology was inadequate to meet the increasing demand for international communications. The satellite was an immediate, substantial improvement, providing six hundred telephone channels (compared to thirty for the cable) as well as a new capacity to transmit television.

Telstar differed in one crucial respect from every other period space-age project: Its funding came predominantly from a private corporation, AT&T, but one with a special status, operating as a U.S. government–sanctioned monopoly. AT&T, as presented in its own promotional material, saw Telstar as "a tribute to the American free enterprise system," and that by "spending millions of dollars of its own money, the Bell System is exploring new voiceways in space to help bring better communications to the nation and the world."

Although Telstar was experimental, the company's investment was aimed at the future, as it looked to create a "worldwide commercial satellite system," capable of "live overseas telecasts—of the Olympics, the Bolshoi Ballet, the pageantry of a king's coronation as well as unlimited circuits for ocean-spanning business and personal telephone calls." Even with its ample resources, AT&T could not do the initial Telstar project alone. The governments of Great Britain and France were partners, providing key resources, as was NASA, which provided the launch services. The effort was seen as an exemplar of private initiative working hand in hand with the public sector.

LEFT: Until Telstar's success, the curvature of the Earth was a significant impediment to long-distance communications. Ground-based towers positioned in series, shortwave radio bounced off Earth's ionosphere, or an undersea cable could partially address the challenge, but they were either cumbersome, unreliable, or limited in communications capacity. Telstar literally rose above the curvature problem, providing easier and greatly increased communications capacity. *AT&T Archives and History Center*

RIGHT: The Andover ground station included, in addition to the massive horn antenna, a control room from which engineers directed the antenna, sent communications to and from space, tracked the satellite, and monitored its operations. Here, the Bell Laboratories' technical team tracks Telstar in the hours after launch on July 10, 1962. *AT&T Archives and History Center*

THE TECHNOLOGY BEHIND TELSTAR

One of the technical marvels of the project was this giant horn antenna, located at Andover, Maine, used to track and communicate with the satellite. Long since dismantled, the antenna measured 177 feet long and weighed 380 tons. Despite its size and weight, the antenna could track the fast-moving, 34½-inch satellite to an accuracy of one-twentieth of a degree. The entire antenna is covered in a Dacron and synthetic rubber structure (called a radome) to protect it from wind, icing, and temperature changes. *AT&T Archives and History Center*

Though Telstar burst into public attention in dramatic fashion in July 1962, the technologies that enabled its success were researched and developed after World War II. The transistor, invented in 1947 at Bell Telephone Laboratories, was crucial. Its small size and durability allowed the development of more compact, complex electronic systems—including the 34½-inch-diameter Telstar.

The satellite depended on the use of microwaves (which were shorter, and thus of higher frequency, than the waves used in commercial radio broadcasts). AT&T had a long-standing interest in microwaves, as they could be transmitted in straight lines and carry more information. As a supplement to its telephone network, AT&T had built a system of microwave ground towers and transmitters throughout the United States. In the years before Telstar, AT&T partnered with NASA on Project Echo, an effort to place large, reflective balloons in Earth's orbit as a means to test the behavior of microwaves in the atmosphere and space, as well as to develop the ground infrastructure for space-based communications. The first Echo was launched in August 1960, nearly two years before Telstar, giving AT&T valuable experience.

This prior experience helps explain one of the most interesting aspects of the Telstar effort—the giant horn antenna that AT&T built at Andover, Maine, used to track the satellite and communicate back and forth. Through its research, AT&T determined that a horn-type antenna was the best design for gathering the narrow beam typical of microwaves. But because Telstar produced such a faint communications signal, the horn needed to be gigantic: 177 feet long and 380 tons. Not only that, to catch the microwaves, the opening of the horn had to be pointed at the satellite with high precision, requiring that the base of the horn be designed for extremely refined control. The antenna was a technical marvel, and a duplicate was built at the French ground station in Pleumeur-Bodou.

Within the context of the Cold War, this collaboration was both a boon and a source of tension. As the fruit of an American corporation, Telstar stood for the defining place of private enterprise in American life, as an alternative to Soviet-style communism. But the importance of satellite communications as a preeminent national concern in the Cold War—as technology and symbol—already had been highlighted in President Kennedy's well-known May 1961 "Moon" speech in which he called for "accelerating the use of space satellites for worldwide communications" as an essential component of the U.S. response to the Soviet challenge. Such a geopolitical perspective intensified in August 1962. After multiple Telstar "firsts," the Soviets announced their plans (presented in the *Washington Post* as "Reds plan own

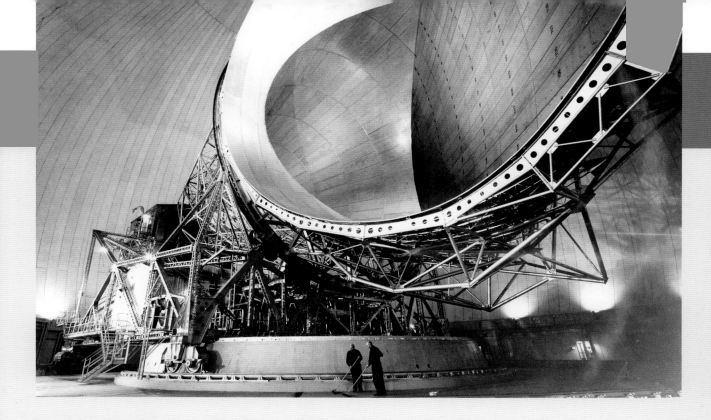

Telstar was the first true active satellite—one that could receive a signal from the ground, amplify it, and then immediately retransmit it. Such a capability set the technical model for live communications, whether via telephone or for television, which technicians, broadcasters, and the American public had come to expect. The critical component was the amplifier—a traveling wave tube—that could take a signal weakened after traveling thousands of miles, boost it (ten billion times, in the case of Telstar), and then retransmit it at a strength that could be picked up and used on the ground.

The combination of these and other technical innovations made Telstar a cultural phenomenon in the summer of 1962 and the harbinger of satellite communications' profound impact in the years to follow.

'Telstar'") to "send up two communications satellites to beam propaganda programs around the world."

As the United States and USSR sought to persuade peoples around the world to align with their respective political values, the question of who should control this powerful communications tool internationally took on high significance. Should this capability be primarily in private hands, or should it be considered an instrument of government policy? Telstar, inadvertently, brought additional scrutiny to AT&T's special place in American life. *Look* magazine ran a fourteen-page spread on the company: "Without AT&T Americans would not be able to call a doctor, watch a nationwide TV show, or fight a nuclear war! . . . read about this corporate colossus.

THE IDEA OF THE FUTURE

COMMUNICATIONS SATELLITES AND ARTHUR C. CLARKE

Arthur C. Clarke is perhaps best known for his collaboration with director Stanley Kubrick on the film classic *2001: A Space Odyssey* (seen here on set). During his long career, he authored nearly one hundred books, primarily on science fiction, science fact, and forecasts of the future. As an early conceptualizer of communications satellites, he maintained a keen interest in their development and future prospects. © *Bettmann/CORBIS*

In 1953, L. P. Hartley opened his novel *The Go-Between* with the memorable sentence "The past is a foreign country: they do things differently there." Yet at the very time Hartley penned those words, the *future*, at least as seen through the lens of science and technology, had become—and would continue to be—a familiar presence. Imaginings of the future, of course, were common in prior decades, but after World War II such reflections took on a new quality and gained wider cultural purchase, spurred primarily by Cold War–funded advances in science and technology and by an increased interest in science fact and fiction as a theme in literature, film, and popular magazines. The idea of probing and understanding the changes that might come in the near and far future gained new credibility—whether to detail the potential horrors of nuclear war or the beneficial possibilities of spaceflight.

Perhaps no single individual thrived in this invigorated climate of the future more than Arthur C. Clarke, preeminent science fiction author and popularizer of science and technology (especially as it related to outer space). As a futurist, he took threads of the present and spun them into stories of what might be. By the time of Telstar, Clarke already was well-known, not least for a brief essay he published in 1945 in which he discussed using space-based satellites to enable intercontinental communications. With Telstar's success, and with other communications satellites already planned for launch, Clarke, in an article for *Life* magazine in September 1964, sought to sketch out the meaning of these developments.

Clarke offered—as did other of his contemporaries—that these first successes in satellite communications augured "a high-quality *global* telephone service, so that any two people on earth can speak to each other at any time of day or night" and that "there will be a rapid development of world-wide TV, so that major events in any part of the planet can be witnessed, live, over the whole globe." But Clarke, with prior experience in such forecasting, cautioned "the most important results of any new technological breakthrough are precisely those that are *not* obvious." Indeed, in 1945 Clarke had assumed that communications satellites would be human-staffed, not the remotely controlled machines they became due to the vast advances in electronics after World War II.

Who owns it? Who runs it? How did it launch Telstar—the world's first privately owned satellite?" Such characterizations only highlighted the stakes in placing the responsibility for satellite communications with either business or government.

But in 1962, policymakers saw the line separating private and public activities through a different lens than they do today. Economic domains such as commercial air travel, radio, and television, and AT&T's federally approved monopoly on telephone service, were highly regulated. Such regulation reflected the view that in

Clarke, thus, looked to tease out "not obvious" ramifications. In retrospect, it is fascinating to see how his envisioning of technological and social change overlaps with our recent history. His underlying premise—that communications technologies diminish the importance of distance in our interactions—is commonplace today, but its realization has played out somewhat differently. "When any two points on the earth's surface can be linked in a second," he surmised, "there will be a steady decentralization of human affairs, and a great reduction in traveling (except for pleasure)." Businessmen, for example, might "live where they choose . . . to run a large law firm or an advertising agency from Bali or Tahiti." The impact of such decentralization on humanity's fundamental social invention, the city, might be "catastrophic," but yet might align with new realities, for the city "may already have exceeded its usefulness . . . as anyone who has seen a Fifth Avenue traffic jam or a Harlem slum will agree." Why, after all, would "men [need] to meet physically when they [could] meet mentally"? Such possibility, he thought, will "ultimately destroy the main need for the city," allowing "mankind to revert to a more natural mode of life"—a view that foreshadowed the sentiments of the emerging environmental movement of the 1960s and 1970s.

His thoughts touched closer to our present experience when he considered the "language problem" posed by global communications. How would we understand each other, given our 'thousands of tongues'? He saw communications satellites as a means to "lift this curse of Babel," first, as creating a shared experience through seeing the same images, but also through the gradual spread of a dominant language in international affairs. "This does not mean that national languages will disappear, but . . . one international language will become universal . . . almost certainly it will be English—with all the cultural, social, political implications that this involves."

Clarke's reflections, though largely presented as *consequences* of communications satellites, were attempts to understand the larger patterns of change of which these satellites were a part—a world that in the 1960s was already moving toward an intensification of global connections in trade and cultural exchange, through international institutions such as the United Nations and via the geopolitics of the Cold War. In this larger context, he was right: communications satellites have played a critical role in creating the global world we know today.

some areas governmental interest was paramount to private interest, a perspective that colored policy discussions on the future of satellite communications. Through the summer of 1962, as Telstar demonstrated its prowess in television broadcasting and other communications feats, Congress debated this problem, eventually approving a "split the baby" solution. AT&T lost out on creating a satellite system under its control, but it became part-owner of a new, government-chartered, quasi-private firm, the Communications Satellite Corporation (Comsat), which in turn was the U.S. representative to a companion international treaty organization, Intelsat. Intelsat, with the involvement of multiple countries, became the primary means of developing communications satellites for the next two decades. In 1969, the fledgling organization, with a small fleet of three satellites, brought the Apollo Moon landing to hundreds of millions of viewers around the world.

Though Washington policy battles can be dry, the debates reflected a new reality. Television had become a critical part of American culture. In 1950, about six million televisions were in use in the United States; after 1960, the number was well above sixty million. Sending images directly into homes, television was a uniquely potent medium—either to promote awareness and critical thinking or to undermine such communal virtues through escapist entertainment. In 1961, Newton Minow, head of the Federal Communications Commission, leaned toward the latter assessment and famously remarked that television programming was a "vast wasteland." Telstar added another dimension to this "promise and peril" perception of the role of television, expanding its reach from primarily national to international markets and bringing events from around the world "live" into everyday experience.

Marshall McLuhan already had coined the term "global village" to characterize this emerging change in human affairs. As Telstar began its broadcasts, famed television anchor Walter Cronkite noted that the satellite makes the "White House and the Kremlin no farther apart than the speed of light"—a sentiment that highlights part of the reason for the satellite's broad impact on the imagination of Americans in 1962. U.S. and USSR long-range missiles were just entering service, making mutual nuclear destruction possible within thirty minutes. For the optimistic, satellite communication—and its promise of instantaneous connection—seemed like an antidote, promoting understanding across cultures and defusing misunderstandings. Writer Arthur C. Clarke, widely regarded as the father of the idea of communications satellites, suggested that the satellite might augur "the rapid unification of the world as one cultural entity, for good or bad." For some, the bad seemed likely to predominate, only accelerating the spread of television's perceived lowest-common-denominator programming tendencies, doing little to nothing to address the era's urgent problems.

But such contextual aspects of the Cold War could become more immediate—and did. On July 9, a day before Telstar's launch, the United States conducted the nuclear-weapons test Starfish Prime, which involved launching and detonating a 1.4-megaton bomb 250 miles above Earth. In Hawaii, nine hundred miles from the North Pacific launch site, the sky was suffused with a dramatic orange-red glow, and microwave communications on the islands were disrupted. The blast immediately increased the intensity of radiation in the lower Van Allen Belt, posing an enhanced risk to Telstar's electronics, eventually leading to the satellite's demise in early 1963. But with the test occurring in such close juxtaposition to Telstar's own placement into space, symbolic comparisons were inevitable. A columnist for the *Boston Globe* characterized the two events as "terror versus Telstar" and noted that "by one of those coincidences only too common in this Alice-in-Wonderland age of nuclear international lunacy . . . modern science tests space communications disruption tactics above the Pacific . . . [while]

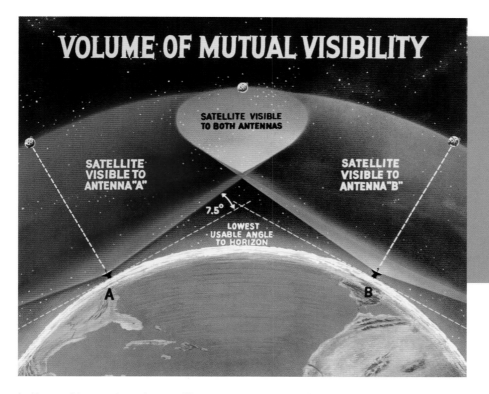

VOLUME OF MUTUAL VISIBILITY

SATELLITE VISIBLE
TO BOTH ANTENNAS

SATELLITE
VISIBLE TO
ANTENNA "A"

SATELLITE
VISIBLE TO
ANTENNA "B"

7.5°

LOWEST
USABLE ANGLE
TO HORIZON

A

B

Telstar flew in an elliptical orbit, roughly 3,700 miles above Earth at its highest point and six hundred miles at its lowest. An important consequence was that the satellite periodically and only for a short time (twenty to thirty minutes) could transmit communications (was "visible") between the ground station in the United States and the ground stations in Great Britain and France. *AT&T Archives and History Center*

half a world away it seeks to utilize space communications to bring peoples and governments toward greater cooperation in peace!"

Before and after its launch, Telstar entered into this confusing context—variously conveying excitement and utopian-tinged prospects for the future, world-changing progress, and the tensions and realities of the Cold War and of television as a cultural phenomenon. In the months leading up to the launch, the project received wide coverage in the media, including a multi-page article in *National Geographic* that was headlined "Telephone a Star." It emphasized that Telstar and its successors would be the first space objects that "millions of people will actually use" (referring to telephone calls) and provide "superhighways" for television and other communications. Amid such speculations and crosscurrents, one of the satellite's technical characteristics only seemed to intensify the sense of a changing present and an impending future: During each orbit of more than two and a half hours that passed over the Atlantic, only twenty to thirty minutes could be used for communications. Television could only be done in bursts, heightening scrutiny of the implications of the new capability.

Soon after Telstar's launch early on the morning of July 10, engineers began to test the satellite. In anticipation of success, a large audience of dignitaries, including Vice President Lyndon Johnson, gathered in Washington, D.C., to commemorate the first phone and television transmissions via satellite. The U.S. television networks carried the events. Initial expectations were that these historic transmissions would be between Andover, Maine, and Washington, D.C.—a strictly U.S. occasion. The very first television transmission was a panning shot of the American flag waving

ABOVE, LEFT: The first test attempted with Telstar was a telephone call between Vice President Lyndon Johnson and AT&T Chairman Frederick Kappel—an indication of the satellite's tremendous political and public relations significance in the Cold War contest with the Soviet Union. Mr. Kappel's voice was sent from the Andover ground station to Telstar, which retransmitted it back to Andover. The call then was sent via land line to the vice president in Washington, D.C. *AT&T Archives and History Center*

ABOVE, RIGHT: Though Telstar had a robust capability to transmit telephone calls, its most notable advance was to transmit television. Soon after achieving orbit, Telstar enabled the first television picture sent to space and back—the United States flag flying in front of the Andover ground station's radome, accompanied by the playing of "The Star-Spangled Banner." This signal went only from the Andover antenna to the satellite and then back to the ground. But soon thereafter, another television transmission from Andover reached the French station at Pleumeur-Bodou. *AT&T Archives and History Center*

BELOW: After Telstar's launch, on July 10, it took several orbits to calibrate and test communications between the Andover ground station and the satellite. Initial transmissions were between only Telstar and Andover, but the satellite then became visible to the ground station in France as a taped TV talk by AT&T Chairman Frederick Kappel aired (seen on monitoring screen in background). Eugene O'Neill, project director for Telstar, displays the "thumbs up" for success. *AT&T Archives and History Center*

in front of the Andover ground station as "The Star-Spangled Banner" played in the background. But the transmission that immediately followed—a conversation among AT&T officials—was picked up at the Pleumeur-Bodou station in France, thereby inaugurating transatlantic television.

On July 12 (still July 11 in the United States), the French ran a test from their side. But rather than broadcast bland panoramas or discussions among dignitaries, they ran a tape of singer Yves Montand and the sights of Paris. A little later in the day, the British finally joined the fun and did a live broadcast featuring the engineers and technicians of Goonhilly. Both east-to-west transmissions, each just several minutes in duration, were carried by American networks and into American homes. As a national marketing event, the French choice seemed more astute. In a U.S.-celebrity name-recognition survey taken in the weeks after, Mr. Montand rose to number three, trailing movie stars Janet Leigh and Kim Novak but ranking above Elizabeth Taylor and Richard Burton. As reported in one magazine, a New York City woman exclaimed, "Just imagine! I was so thrilled! Yves Montand! Live! Straight from Paris!"—indicating television's propensity to confuse recorded and live events. CBS, referring to its coverage of the Goonhilly broadcast, proclaimed "it was only eight minutes long . . . it had no stars, no script . . . it went on the air at 3:22 a.m. British summer time . . . it was one of the most important television broadcasts ever presented," and it was brought to American viewers courtesy of the network. Telstar had arrived. Its coverage on TV and in newspapers resembled, according to one observer, a "space fever chart."

With these transmissions as preview, U.S. and European television networks moved from opportunistic presentations to a coordinated, planned transatlantic extravaganza, set for July 23. Several hundred million viewers on both sides of the Atlantic, in sixteen nations (including communist Yugoslavia), watched the telecast, which began with a split-screen image of the Eiffel Tower and the Statue of Liberty, then an excited exhortation of "Go, America, Go" as the U.S. portion of the program started. It was a combination of seriousness—presenting a portion of President Kennedy's press conference, as he talked about monetary policy and the dangers posed by USSR and U.S. nuclear testing—and travelogue, showing scenes of iconic American locales, such as Mount Rushmore, accompanied by singers belting out the "Battle Hymn of the Republic."

The presentation was a genre already well-established in television programming known as "Wide Wide World," a format that sought to bring the distant, the exotic, or merely the unfamiliar into American and European homes, making it at least partially familiar to an average viewer. The European portion of the show, which occurred a couple of hours later on a subsequent orbit of the satellite, was much the same, completing what was widely reported as "a historic achievement, a notable victory for the West in its space and communications race with the Soviet Union."

This series of July broadcasts, from July 10 to 23, encapsulated the uncertainties of 1962 and the possible role of transcontinental television in the Cold War and day-to-day life. Several questions motivated political leaders in this context, infusing Telstar with specific meaning. Should television via satellite, with its broad geographic reach, emphasize high-minded news coverage of political import? Or emphasize the projection of idealized concepts of the nation—whether of the United States, France, Britain, or others? Or reflect television's preeminent role in conveying popular entertainment (which some regarded as "vast wasteland")? All of this invoked the key question: Who would decide?

In the months to follow, as Telstar undertook additional broadcasts, all of these perspectives and contentions jostled against each other. The Pope told pilgrims gathered in Rome that Telstar had "helped strengthen brotherhood among peoples" and "marked a new stage of peaceful progress." In turn, when Vice President Johnson visited the Pope in fall 1962, he presented His Holiness with a model of the satellite as a gift. Others weighed in as to whether Telstar's scarce airtime should be used for news or the actual fare of popular entertainment—period shows such as *Gunsmoke*, *The Lone Ranger*, and *Yogi Bear*. This sentiment could turn dark. Political philosopher Ayn Rand saw in Telstar an avenue to totalitarian suppression of free speech, asking, "Which one of us will obtain equal time on that global medium? And if we do not, how will we make ourselves heard?"

Optimistic assessments, though, were more common. Historian Arnold Toynbee penned for the *New York Times* a long essay called "A Message for Mankind from Telstar," arguing that the technical progress represented by the satellite paled in

On the second day of Telstar's operation, July 11, 1962, the French succeeded in transmitting the first television program from Europe to the United States, seen by millions of Americans on their home television sets. The program was prerecorded and included a musical performance by Yves Montand, a renowned French singer (pictured here). *AT&T Archives and History Center*

While Americans were justly proud of Telstar and its accomplishments, so were the French. A critical element of the Telstar system was the ground station at Pleumeur-Bodou in western France, which facilitated the sharing of Telstar communications with the whole of Europe. To commemorate their contribution, the French issued this stamp, highlighting their role in the very first transatlantic television transmissions, as well as another stamp that depicted the Pleumeur-Bodou station. *Smithsonian National Postal Museum*

On July 23, 1962, American and European officials organized via Telstar a television extravaganza for viewing publics on both sides of the Atlantic. Programming from the United States came first; it included a major-league baseball game and part of a White House press conference by President Kennedy—all packed into about twenty minutes. Here, Londoners in a pub watch the President "live" (hazily visible on the television screen).
© Bettmann/CORBIS

NO DRINKS WILL BE SERVED AFTER SECOND BELL NOT EVEN OFF SALES

comparison to its "new hope for the survival of the human race" against the threats of nuclear annihilation. The reason, he offered, lay in the power of television, which was "the nearest thing to meeting physically face-to-face," and once that process of communication began, it offered the hope of "growing together in a single family," erasing the divisions that might lead to war. But such high purpose required a particular kind of programming: "television must, in fact, be documentary . . . it must display to each section of the human race a fair sample of all the other sections' ways of life." The worry for Toynbee was the predominance of "recreational" television, which in the Telstar-age of international communications threatened to "breed mutual contempt [among nations] instead of the mutual esteem and affection and confidence we so sorely need."

A few days before Toynbee's essay appeared, Telstar's scarce time was used to broadcast to France a twenty-minute program on Marilyn Monroe's death, which included "pictures of Miss Monroe's secluded home and an outside view of the bedroom where her nude body was discovered." Such programming ran in a direction opposite to Toynbee prescriptions for the new technology. A period culture watcher mused that Telstar might be introducing the "Age of Ephemera—the one day sensation, wowing 'em simultaneously in Paris, Peoria, Pretoria, and Peru," a trend that artist Andy Warhol reduced to fifteen minutes a few years later. Paris fashion designers, enamored with Telstar's symbolism, shared their sketches for upcoming fall fashions with Americans via the satellite, breaking a long-standing tradition of embargoing designs until their display on the runways of the City of Lights. At the House of Dior,

viewers saw a designer "marshaling a parade of the new silhouette, the 'Arrow Look'" and were informed that this look behind the scenes was permitted "in the interests of science."

Though the question of television predominated in debates and discussions of Telstar, the satellite also performed numerous tests of telephone technology, faxes, and high-speed data transmission. By the end of October, it had made more than 650 tests and demonstrations of all kinds. These included a conference call among local officials from twenty-three U.S. and twenty-three European cities; the synchronizing of master clocks in the United States and England; and the placing of an order for flowers from Minneapolis to be delivered in Paris.

Almost all these various activities—from the grand and serious to the mundane—were reported by major newspapers and magazines, making Telstar "tops in capture of interest." Models of the satellite travelled around the U.S. to museums, schools, and local community groups. Not surprisingly, in the energetic consumer culture of the early 1960s, this fascination translated into a boom in satellite-themed trinkets: cigarette lighters, thermoses, tie clips, stamps, and more. In August, the British rock group Tornados released the instrumental "Telstar," replete with thin, electronic modulations evocative of the space age. The record became a hit in America and Europe, reaching No. 1 on the U.S. pop charts, making the Tornados the first British group to achieve that distinction.

In the yin and yang of Telstar, a mirror of the state of television and of the broader culture of the Cold War era, popular frivolity and serious politics shared transatlantic airtime, providing opportunities to see and be seen in new ways, to reimagine the relations among faraway places and events and what was happening at home. As African-American leaders, in the midst of the civil rights movement, poignantly stated, the satellite could be used to share their struggle on an international stage and gain new voices of support: "The whole world knows what's happening here. The whole world is watching. . . ." Though that last phrase was not quite true, and carried different meanings for different people, it captured the essence of what Telstar brought into the world of 1962—and of what would follow in the years after as global communications, via satellite and undersea fiber-optic cable, gradually became part of the fabric of everyday life.

—*Martin Collins*

TELSTAR
SPECIFICATIONS

MANUFACTURER:
Bell Telephone Laboratories
DIAMETER: 34.5 in. (87.6 cm)
WEIGHT: 170 lbs. (77.1 kg)
LAUNCH VEHICLE: Delta
ORBIT INCLINATION: 45°
APOGEE: 3,687 mi. (5,933 km)
PERIGEE: 592 mi. (952 km)

THE KH-4B WAS the last and most advanced camera used in America's first successful photoreconnaissance satellite program. Codenamed CORONA, the program launched 145 satellites with cameras and film from January 1959 to May 1972. This equipment acquired photographs of the Soviet Union and other important targets around the world, for intelligence purposes.

CORONA's development proceeded slowly as personnel at the Department of Defense (DOD), Central Intelligence Agency (CIA), and independent contractors overcame numerous technical challenges. After twelve consecutive failures and one successful test flight, the first mission that returned exposed film took place in August 1960. Yet many failures occurred after. However, officials resolved these technical problems, and by the mid-1960s the CORONA program was functioning smoothly. More advanced cameras and other improvements greatly increased the amount and quality of the photography and enabled the satellites to remain in orbit much longer.

CORONA was not the only U.S. photoreconnaissance satellite program implemented during this period. The U.S. intelligence agencies began launching GAMBIT satellites in 1963 to acquire higher-resolution photographs. However, these cameras only covered a limited number of targets, and CORONA was the only system capable of imaging large areas of the Earth. CORONA's enormous intelligence value helped stabilize international relations during the dangerous Cold War period.

CORONA KH-4B /3
CAMERA

Like the other reconnaissance satellite programs, CORONA was shrouded in secrecy. Only a limited number of government officials knew of its existence and had access to its photography and the information it revealed. With the end of the Cold War, the government began examining whether any obsolete reconnaissance satellite programs could be declassified so that the public could learn about their immense contributions to national security. President Bill Clinton mandated in 1995 that the CORONA program be declassified and the photographs and surviving hardware be released to the public. The National Air and Space Museum quickly obtained the KH-4B camera, which had originally been assembled in 1974 from the program's remaining spare parts.

CORONA's Origins

Under great secrecy, the U.S. government initiated several projects in the 1950s to acquire timely and accurate information about the USSR's military forces. The new

The Museum's KH-4B camera on display at the National Mall Building. Built from spare parts two years after the program ended, it is the only surviving camera from the CORONA program. *NASM*

Richard Bissell Jr., Deputy Director of Plans at the CIA. Bissell headed the CORONA program from its beginning in 1958 until he left the agency in 1962. *NRO*

intercontinental ballistic missiles (ICBMs), capable of striking the United States with nuclear warheads, were of particular concern. Collecting intelligence on America's primary Cold War adversary was extremely difficult; the Soviet Union was a huge country that spanned eleven time zones, with a government that imposed tight security on virtually all of its citizens' activities. U.S. ground stations in neighboring countries, as well as U.S. ships and planes operating along Soviet borders, provided significant but limited information.

Scientific advisors to the White House, CIA, and military advocated development of a high-altitude aircraft to overfly the USSR and acquire the critical intelligence needed until better, less vulnerable satellites were built. These remarkable individuals included Dr. William Baker of Bell Labs; Merton Davies and Amrom Katz of RAND; Dr. Sidney Drell of Stanford University; Dr. Richard Garwin and Dr. James Killian of MIT; Dr. Edwin Land of Polaroid; and Frank Lehan, Dr. William Perry, and Dr. Edward Purcell of Harvard. They not only contributed greatly to the nation's reconnaissance programs in the 1950s but for many years after.

The CIA created the U-2 spy plane as the short-term solution called for by these scientific advisors. From 1956 to 1960, the U-2 conducted twenty-four missions over the USSR and brought back photographs of approximately 15 percent of the country, providing important information on airfields, atomic energy installations, the rocket launch complex at Tyuratam, and other key facilities. However, these flights were extremely provocative, and the Soviets vigorously protested them in private. When the Soviets shot down Gary Powers and his U-2 in May 1960, all overflights of the USSR ended. Fortunately, the CORONA satellites were about to enter service.

At the same time that the U-2 became operational in 1956, the air force initiated the WS-117L program to develop the satellites that the scientific advisors had recommended. The most promising photographic system involved sending a camera into orbit and returning the exposed film in film return capsules. To ensure greater security and faster implementation, President Dwight Eisenhower ordered in early 1958 that this photographic system be severed from the WS-117L program and jointly managed by the CIA and air force. Richard Bissell Jr., Deputy Director for Plans at the CIA, headed the project, codenamed CORONA.

The major contractors were Itek (camera), Eastman Kodak (film), General Electric (film return capsule), and Lockheed Missiles and Space (Agena satellite carrying the camera, film, and film return capsule). These firms shipped their components to the Advanced Projects facility in Sunnyvale, California, where the components underwent extensive testing and installation in the Agena satellite. The satellite was then trucked to Vandenberg Air Force Base in California for additional testing before being declared ready for launch.

Modified U.S. Air Force Thor intermediate-range ballistic missiles, assisted by the main engines in the Agena upper-stage vehicles, boosted the satellites into low-Earth, near-polar orbits to enable complete coverage of the USSR's territory. The cameras operated only upon command so that the limited amount of film would not be wasted. Once all the exposed film was wound onto the reels in the film return capsule, the capsule separated from the satellite, fired its small rocket, and

A CORONA launch from Vandenberg Air Force Base in California. Modified U.S. Air Force Thor intermediate-range ballistic missiles boosted the Agena satellite carrying the camera, film, and other equipment into orbit. *NRO*

reentered Earth's atmosphere. Protected by a heat shield, the capsule deployed a parachute at about sixty thousand feet and was then snagged midair by air force planes in the recovery area northwest of Hawaii. In the event that air recovery failed, navy divers would retrieve it from the ocean's surface. The film was removed in Hawaii and flown to the East Coast for processing, and then to Washington, D.C., for analysis. Because the launches at Vandenberg could not be hidden from the public and there would be inquiries about the payloads, the government gave the project the cover name of DISCOVERER and explained that it was testing environmental conditions in space.

All of the program's missions from January 1959 through June 1960 failed for various reasons. On August 11, 1960, DISCOVERER XIII finally reached orbit, and navy divers subsequently recovered its capsule from the water. This was the first object ever retrieved after orbiting the Earth. The satellite did not carry a camera or film; instead it held diagnostic equipment to help determine the causes of the long string of failures.

One week later, DISCOVERER XIV became the first successful mission to carry a camera and film. After seventeen revolutions around the Earth in a little more than twenty-four hours, the satellite ejected the film return capsule. An Air Force

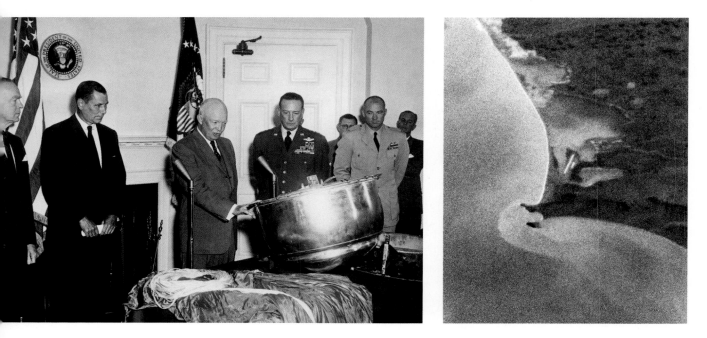

LEFT: President Dwight Eisenhower inspects the DISCOVERER XIII capsule at the White House shortly after its recovery northwest of Hawaii in August 1960. The first successful mission in the CORONA program, it carried diagnostic equipment. *National Park Service*

RIGHT: Taken on August 18, 1960, this photograph of the Mys Shmidta airfield in the northern Soviet Union was the very first acquired by CORONA. *NRO*

C-119 snagged its parachute and reeled it in. In just seven passes over the USSR, the satellite had photographed more of that nation than had been accomplished in all twenty-four U-2 overflights. However, the maximum ground resolution of thirty-five feet (the smallest object that could be identified in the photographs) was much poorer than that obtained by the U-2 cameras.

Only nine capsules were recovered with usable photography from the next twenty-seven launches, through the end of 1961. Problems with the Agena satellites or with the cameras and film were common. Nevertheless, the intelligence value of the returned imagery was immense, as it captured increasing numbers of Soviet targets.

Improvements to CORONA

CORONA incorporated several important technical advances in the early 1960s. The C Triple Prime camera (KH-3), which first flew in August 1961, produced a ground resolution as high as twenty feet. A more capable Agena satellite enabled longer missions with more passes over the USSR. DISCOVERER XXXVIII, the last flight with a single C Triple Prime Camera, took place in January 1962. It marked the end of the DISCOVERER cover story, which no longer served any useful purpose. The authorities released only very limited information regarding subsequent CORONA launches.

In February 1962, the MURAL camera (KH-4) entered service. It consisted of two C Triple Prime Cameras, one looking 15° aft and the other 15° forward. Simultaneous operation of the two cameras produced stereo photography in which the two photographs of the same site or object would appear to be three-dimensional when examined by a photo analyst through a special viewer. This made it easier to determine the size of targets such as buildings or missiles. The MURAL system carried twice the amount of film and also achieved a maximum ground resolution of

THE DEFENSE METEOROLOGICAL SATELLITE PROGRAM

One of the biggest challenges in CORONA was determining the weather over the targets when the cameras were above to photograph them. Because of the small film load and the fact that most of Earth's surface did not have any targets of interest, the cameras did not operate continuously, but only when air force ground controllers commanded them to do so. As with all optical systems, however, CORONA could not image through clouds.

Timely and accurate weather data was required, so if a target was cloud-covered, controllers could direct the cameras not to photograph it at that time. The lack of such information was an enormous problem with CORONA's early missions. Approximately half of the imagery it produced was of cloud cover.

The National Reconnaissance Office (NRO), which operated CORONA and other reconnaissance satellites, initially hoped to rely on NASA's TIROS weather satellites for the requisite data. However, these satellites were unable to provide it. The NRO began developing a classified weather satellite system in 1961. Designated the Defense Meteorological Satellite Program (DMSP), the goal was to have two spacecraft in polar orbits at all times. With cameras and other sensors, one would gather weather data over the Eurasian landmass at about 7:00 a.m. local time and the other at approximately 11:00 a.m. local time. The satellites would downlink the data to one of two ground stations in the United States when within range and, after being processed, this weather data would be sent to the NRO for use in commanding CORONA and other photoreconnaissance systems.

Although many of the early launches failed, the single satellites that occasionally reached orbit sent back data that helped CORONA and other photoreconnaissance satellites acquire more useful imagery. The NRO finally placed two DMSP spacecraft in orbit at the same time in 1965 and launched replacement satellites to maintain this constellation. This resulted in a substantial increase in the quality of the photography. The air force assumed management of DMSP later in the 1960s, but its primary mission of providing timely and accurate weather data for photoreconnaissance satellites did not change.

DMSP remained a classified program until 1973. Thereafter, all of the data its cameras and other sensors acquired was freely shared with the public.

---

ten feet. Small stellar and index cameras were added to the MURAL system, which imaged the stars and Earth when the main cameras operated. This photography, when coupled with the satellite's orbital data, greatly assisted photo analysts in determining exactly what part of the Earth the main cameras had imaged.

Of the twenty-six MURAL systems launched, twenty-four reached orbit. Twenty film return capsules were recovered from February 1962 through its last flight in

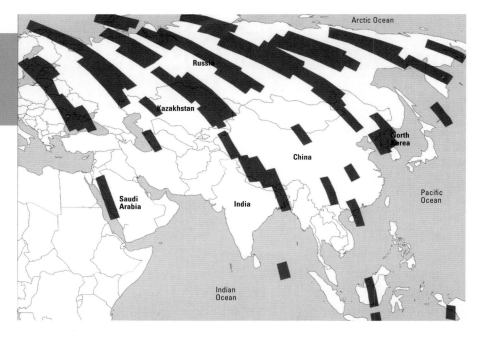

This map depicts the Eurasian areas photographed by a CORONA mission in the mid-1960s. *NRO*

December 1963. The longest mission was a little more than five days, and the quality of the imagery was generally excellent.

The next CORONA camera was the J-1 (KH-4A). It finally satisfied the requirement to carry a large amount of film in order to reduce the number of launches needed to obtain the required photographic coverage, and thus decrease the costs and risks. The J-1 was similar to the MURAL, but it carried twice the amount of film and a second film return capsule. The first was jettisoned and then recovered after half the film had been exposed, and the second was then filled, to be recovered later. To extend mission life, the cameras could be deactivated for up to twenty-one days after the first capsule was ejected. Three solid-propellant strap-on rockets were added to the Thor launch vehicle to enable it to place the heavier payload into orbit.

Fifty-two J-1s were launched between September 1963 and September 1969, with recovery of ninety-four of the 104 film return capsules. The longest flight was sixteen days. A single mission could image eighteen million square miles of the Earth in stereo photography, almost four times as much as earlier in the program. The cameras routinely produced ten-foot ground resolution and, at times, achieved better than seven feet.

The KH-4B and the End of the CORONA Program

Program officials began development of a new dual constant rotator camera system in 1965 to further improve ground resolution and camera flexibility. Designated the J-3 or KH-4B, the new system could accommodate a variety of film types and operate in orbits as low as ninety-two miles. The air force launched the first in September 1967, followed by seven more by the end of 1969. The longest flight was eleven days, and the highest ground resolution achieved was about five feet.

At the beginning of 1969, fifteen CORONAs (three KH-4As and twelve KH-4Bs) remained in the inventory. The initial launch of CORONA's successor, codenamed

HEXAGON, was scheduled for late 1970. It carried far more film and four film return capsules, and its cameras achieved a higher resolution. However, HEXAGON faced considerable development problems, and meeting the late 1970 launch date was increasingly unlikely. Officials had to decide whether to procure more CORONAs and, if so, how many, to guarantee coverage in the event of HEXAGON delays. By the beginning of 1970, they had concluded that there was a 95-percent probability of launching the first HEXAGON no later than June 1971. All the remaining CORONAs should thus be launched, and there was no need to procure more. The program managers also decided to place at least two into orbit after June 1971 to provide coverage in case the first HEXAGON failed.

This schedule guided the remaining CORONA flights. Four KH-4Bs were launched in 1970, three in 1971, and the last two in early 1972. All were successful, except for the February 1971 mission in which the Thor booster exploded shortly after liftoff. The longest flight was eleven days, and the maximum ground resolution continued to be five feet.

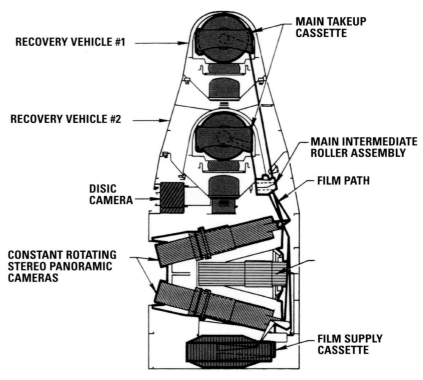

RECOVERY VEHICLE #1

RECOVERY VEHICLE #2

DISIC CAMERA

CONSTANT ROTATING STEREO PANORAMIC CAMERAS

MAIN TAKEUP CASSETTE

MAIN INTERMEDIATE ROLLER ASSEMBLY

FILM PATH

FILM SUPPLY CASSETTE

Major Components of the J-3 System

The J-3 (KH-4B) was the last and most advanced camera used in CORONA. It could detect objects as small as five feet on a side, a vast improvement over earlier cameras. *NRO*

CORONA's Intelligence Value

From January 1959 to May 1972, 145 CORONA satellites were launched and 165 film return capsules were recovered. The system had imaged more than one billion square miles of the Earth's surface at increasingly higher resolutions.

The importance of this photography to U.S. civilian and military policymakers cannot be overstated. With respect to the Soviet Union—the most critical target throughout the program—the system detected and photographed all medium-range, intermediate-range, and ICBM complexes. The first few successful missions quickly ended the acrimonious "missile gap" debate in the late 1950s and early 1960s. Many influential civilian and military officials had claimed that the USSR was planning to deploy thousands of ICBMs and the United States was planning to deploy up to ten thousand such missiles as a counterweight. The imagery from these flights proved that the Soviet Union had no more than twenty-five, which enabled the United States to create a far smaller force of one thousand ICBMs as an adequate deterrent.

CORONA continually monitored the buildup of Soviet missile forces, and by mid-1964 it had photographed the last of the twenty-four ICBM complexes. It enabled the intelligence agencies to make definitive statements such as the one contained in a 1968 National Intelligence Estimate that no "new ICBM complexes have been established in the USSR during the past year." These firm conclusions would not have been made without complete confidence that CORONA would have detected any new construction. With its ability to accurately track Soviet missile deployment, the system also provided solid intelligence upon which to enter strategic arms control negotiations in 1968. These negotiations ended with the signing of the first Strategic Arms Limitation Treaty and first Anti-Ballistic Missile Treaty with the USSR in 1972.

CORONA photographed Soviet naval shipyards for ballistic-missile submarines and surface vessels, enabling analysts to determine what warships were under construction and when they joined the fleet. Coverage of aircraft factories and airfields provided updates on the numbers and types of bombers and fighters. The system detected and monitored the anti-ballistic missiles around Moscow and the various radars associated with them, as well as the numerous surface-to-air missile complexes throughout the country. Among other things, imagery of nuclear facilities helped provide advance warning of upcoming tests, estimates of the production of key materials, and weapons storage sites. Photography of biological and chemical warfare production facilities and testing sites assisted in determining the state of these programs. CORONA made important progress in obtaining an accurate inventory of the bases, weapons systems, and support equipment of the Soviet ground forces.

The system also imaged Chinese, North Korean, North Vietnamese, Cuban, and Warsaw Pact airfields, naval shipyards, naval bases, missile complexes, ground forces, and other targets. One of the most important U.S. objectives was determining the scope of the Chinese nuclear program. Photography from CORONA assisted

A June 28, 1962, image of the Yurya intercontinental ballistic missile complex in the USSR. Because of the dangers these weapons posed to the United States, these sites were a top priority target during the entire CORONA program. *NRO*

RECOVERING THE FILM

Returning the exposed film safely back to Earth was a complex and challenging procedure. While in orbit, it wound onto a reel in a film return capsule after being run through the camera. When directed by air force ground controllers at an Alaska station, the capsule separated from the satellite at an altitude of more than one hundred miles above the Earth, fired its retro-rocket, and began its journey to the recovery area northwest of Hawaii.

The capsule was aluminum and weighed a little more than four hundred pounds with the film. An ablative heat shield, made of the same materials that helped to safely bring back space-craft carrying astronauts, was attached to the capsule to protect it from the tremendous heat generated during reentry. A thin layer of gold covered the capsule's exterior to afford additional protection in the event of a heat shield failure. At about sixty thousand feet, automatic systems jettisoned the heat shield and deployed a large parachute.

An Air Force C-119 is about to snag the parachute of a film return capsule. Aerial recovery was the most common method of recovering exposed film from space. *NRO*

These systems also activated a flashing light and radio beacon to assist the recovery forces in tracking and retrieving the capsule. Specially trained air force crews of the 6594th Squadron based in Hawaii had the primary responsibility for retrieval. The pilot maneuvered the aircraft, trailing a long boom with hooks to snag the parachute and reel in the capsule at an altitude of approximately fifteen thousand feet. If aerial recovery failed, divers from navy ships retrieved the capsule after it landed on the surface of the Pacific. To prevent recovery by any foreign ships in the event the divers' efforts were unsuccessful, the capsules had a salt plug that dissolved after three days, sinking them.

Many problems plagued the CORONA program in the early years, and recovery of a capsule with usable film was a major event. This situation began changing dramatically by the time the KH-4A system entered service, carrying two capsules. Air force planes or navy ships recovered forty-three of the fifty-two capsules launched on twenty-six KH-4A missions from August 1963 to September 1969. The record was even more impressive with the follow-on KH-4B. All but two of the thirty-four capsules launched in seventeen missions from September 1967 to May 1972 were recovered.

Art Lundahl, director of the CIA's National Photographic Interpretation Center, headed the CIA's analysis of imagery acquired by aircraft and satellites until his retirement in 1973. *NRO*

The CORONA program was operational long before digital photography was invented. Its film had to first be developed and then examined by highly trained photo analysts using light tables and other equipment, to obtain any intelligence information. Copies needed at other sites had to be made and then securely transported by couriers.

With the start of U-2 operations in 1956, the CIA established its own facility where analysts worked on the photography from that aircraft and, beginning in August 1960, the CORONA satellites. It created the National Photographic Interpretation Center in early 1961, where more than one thousand personnel from the armed services and intelligence agencies examined the imagery returned by these platforms. Art Lundahl, a pioneer in the field of photographic analysis, headed the center until his retirement in 1973.

A single CORONA photograph covered approximately 1,500 square miles. Early missions brought back several hundred photographs, whereas later missions, with their much larger film loads, returned well over a thousand. Immediately after developing the film, personnel performed a quick review and selected the photographs of the best quality and highest possible intelligence interest to be examined first.

One of the key tools analysts used was a light table, which enabled them to view imagery through magnifying scopes. Improved ground resolution during CORONA's lifetime, from roughly forty feet with the first cameras to six feet with the last, allowed the detection and identification of many more objects. Stereo coverage, first obtained in 1962 with the incorporation of two cameras, made the measurement of objects much easier and accurate.

analysts in establishing an inventory of production facilities, following the development and deployment of weapons with which to deliver nuclear warheads, and tracking activities at the nuclear test site.

CORONA frequently photographed the volatile Middle East. In 1967, it was critical in evaluating and supporting Israeli claims regarding the destruction of Egyptian, Syrian, and Jordanian aircraft on the ground. Three years later, CORONA proved to be the only intelligence collection system capable of evaluating Israeli-Egyptian claims regarding compliance with their ceasefire agreement.

CORONA also satisfied other national security requirements, one of the most important of which was mapping. Although not originally designed for this purpose, CORONA's photography soon became an invaluable cartographic tool. U.S. military forces needed accurate maps and associated products for several reasons, including the targeting of long-range missiles and operations in Southeast Asia and other theaters. At the start of the program, most of the world, apart from the United States

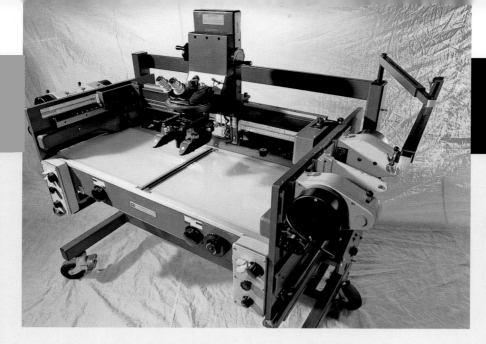

An AIL 1540 light table used to analyze imagery beginning in 1970. The powerful stereoscopic viewer enabled analysts to magnify objects in photographs many times. *NASM*

Analysts soon determined that there were keys to identifying certain targets. SA-2 surface-to-air missile complexes were all built in a Star of David configuration. The Soviets exported large numbers of these weapons, and this fact made it much easier to identify them in such countries as Cuba, Egypt, North Vietnam, and Syria. Intercontinental ballistic missile complexes were always served by rail lines to bring in the necessary construction supplies and missiles. They were also surrounded by multiple security fences.

The workload at the National Photographic Interpretation Center was staggering. It not only analyzed U-2 and CORONA imagery but photography from other platforms such as the SR-71 aircraft and the GAMBIT satellites. Through it all, the facility continually developed new equipment, techniques, and processes to continue providing timely and accurate intelligence to civilian and military policymakers.

and Europe, were poorly mapped. Geographic locations in much of the USSR east of the Urals, for example, were only known to within fifteen to thirty miles. Some places in this vast region could not be located at all. Beginning in 1965 with the J-1 (KH-4A), the intelligence officials managing CORONA determined that mapping and charting needs mandated coverage of one million square miles within the Sino-Soviet Bloc and twenty-three million square miles outside the bloc. As a result, they began assigning a specific percentage of the film on many missions to meet these requirements and directed that a small number of missions be flown dedicated solely for this purpose. CORONA essentially satisfied these objectives for the Sino-Soviet Bloc before the program ended in 1972, but it fell short for the areas outside of it largely because many were covered with clouds much of the time.

CORONA also provided information on the Soviet civilian space program, particularly its manned lunar landing program that competed with Apollo—meeting another national security requirement. After President John Kennedy announced in

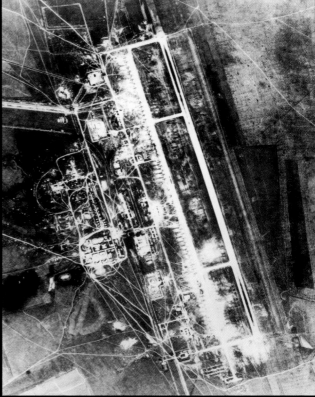

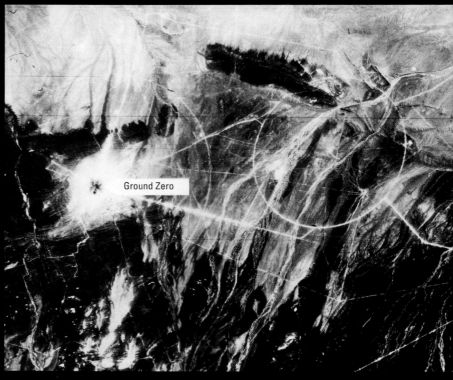

Ground Zero

ABOVE, LEFT: The Severodvinsk naval shipyard in the Soviet Union on February 10, 1969. The ice-breaker tracks on the river indicate that vessels are about to leave. *NRO*

ABOVE, RIGHT: An August 20, 1966, photograph of the Dolon airfield in the USSR. Analysts were able to identify the individual transport and bomber aircraft on it. *NRO*

BELOW: The Chinese nuclear test site at Lop Nur on October 20, 1964. CORONA imagery of this facility allowed the United States to accurately predict the detonation of the first Chinese nuclear device. *NRO*

1961 that the United States would land men on the Moon and return them before the end of the decade, top officials at the White House, National Aeronautics and Space Administration (NASA), and other agencies needed to know whether the USSR was engaging in a similar effort. Just as with the United States, one of the things the Soviet Union needed to do to succeed in this endeavor was to build a massive rocket that would require a new, much larger launch pad. In 1963, CORONA and GAMBIT first photographed such a pad in the early stages of construction at the massive Tyuratam launch complex. Designated Complex J by the CIA, the satellites regularly imaged this site during subsequent missions; their imagery showed that construction was proceeding slowly.

Based on this and other data, a 1967 National Intelligence Estimate concluded that the Soviets could attempt a manned lunar landing in mid-1969 at the earliest but that sometime in 1970 was more likely. Coverage the following year obtained the first images of a gigantic new rocket, which the Soviets designated N-1, at Complex J. However, primarily because that rocket was not flight-tested and construction of Complex J had not been completed, the intelligence agencies concluded in 1968 that the earliest date that the USSR could attempt a manned lunar landing was probably in late 1970 or 1971. In short, it did not have a program that could beat Apollo. CORONA and GAMBIT continued to image Complex J, and all of their photographs reinforced this conclusion. In the end, the United States was the first to land humans on the Moon with the Apollo 11 mission in July 1969; the Soviets never attempted this feat.

CORONA helped meet one more important requirement, unrelated to national security. In the mid-1960s, NASA sought to install sophisticated cameras in its spacecraft to conduct remote sensing of the Earth for use in agriculture, geology, oceanography, urban planning, and other civilian disciplines. However, the intelligence agencies successfully opposed NASA's plans to openly use cameras equivalent in performance to their classified ones and to openly distribute high-quality imagery. Instead, they established a project under which a limited number of scientists from several federal civilian agencies received the necessary security clearances and could access photography from CORONA and other overhead systems. Most of these images were of the United States, and there were only a few instances of access being given to view photographs of foreign countries. The U.S. Geological Survey was the single largest user, and it updated many of its maps of the United States with the classified material. The Office of Emergency Planning, the Agency for International Development, the Department of Agriculture, and the Environmental Science Services Administration also benefitted greatly from the project.

CORONA was developed to provide U.S. civilian and military policymakers with timely and accurate intelligence on the USSR and other nations around the world. Despite numerous problems early in the program, it quickly became a reliable and effective system. The installation of advanced cameras and other items greatly increased the amount and quality of the photography acquired. CORONA was critical to national security during a period of threats and uncertainty.

—*James E. David*

KH-4B
SPECIFICATIONS

MANUFACTURER:
Itek Corporation
TYPE: Stereo/Panoramic
Constant Rotator
LENS: 24-inch focal length,
f/3.5 Petzval design
FILM CAPACITY: 15,750 ft. (4,801m)
of 70mm film per camera
MAXIMUM GROUND RESOLUTION:
6 ft. (1.83m)
COVERAGE OF INDIVIDUAL FRAME:
8.6 × 117 nautical mi.
(15.9km × 216.7km)

Go Baby, Go!

THE JULY 16, 1969, LAUNCH of Apollo 11—the first human mission to land on the Moon—marked the climax of the Cold War space race between the United States and the Soviet Union. NASA, the media, and the public hailed a new era in spaceflight, although more pressing problems at home—social inequality, racial strife, and the protracted war in Vietnam—soon accelerated the space agency's drift from the national spotlight as NASA budgets declined sharply in the 1970s.

Nonetheless, the success of Apollo signaled an important milestone in the evolution of liquid-fuel rocket-engine technology. The first stage of the huge Saturn V that sent American astronauts to the Moon sat atop five F-1 engines, each of which burned for 2.5 minutes and produced 1.5 million pounds of thrust— far surpassing all other rocket-propulsion systems then in existence. In its scale and operation, the F-1 engine expressed NASA's preference for technological stability and continuity rather than radical innovation. It did not at first incorporate disruptive technologies, but rather followed a path of incremental product improvement that captured on a much larger scale the major breakthroughs in liquid-fuel rocket technology dating back to World War II. This approach, however, did not preclude significant innovations, especially in materials research and development that the F-1's size and performance demanded. Moreover, like the U.S. human spaceflight program and the launch vehicles for its missions, the

THE F-1 ENGINE / 4

F-1 owed its operational capabilities and technological attributes to the military— primarily the air force and the army—and the interconnected network of industrial contractors that had spearheaded the development of long-range missiles before NASA's founding in 1958.

Standing at 363 feet tall, the three-stage Saturn V, the brainchild of transplanted German rocket engineer Wernher von Braun—developer of the V-2 ballistic missile—far exceeded in height and weight all U.S. civilian and military launch vehicles. The cluster of five F-1 engines attached to the first stage (S-IC) produced a combined thrust of 7.5 million pounds, a figure that placed the Saturn V—weighing in at 6.2 million pounds fully fueled—in a class by itself. The human lunar mission required such a big scale-up in size and thrust. Engineers at NASA's George C. Marshall Space Flight Center and the Rocketdyne Division of North American Aviation—the firm responsible for developing and producing the F-1—followed, however, a deliberately conservative design strategy to minimize operational risks

The Smithsonian Institution obtained this F-1 engine, built in 1963 by the Rocketdyne Division of North American Aviation, for the National Air and Space Museum in 1970. Along with a one-quarter cutaway, it is located in a mirrored exhibit space to give museum visitors the illusion of looking at all five F-1 engines at the base of a Saturn V launch vehicle. *NASM*

ABOVE: The Apollo 11 Saturn V lifts off from launch complex 39A at Kennedy Space Center, Florida, on July 16, 1969. Astronaut Neil Armstrong stepped onto the lunar surface on July 20. *NASA*

OPPOSITE: Wernher von Braun stands beside the clustered F-1 engines on the Saturn V first stage, on display at the U.S. Space and Rocket Center in Huntsville, Alabama, in the early 1970s. *NASA*

in favor of simplicity, durability, reliability, and safety.

It had not always been this way. Work on what became the F-1 engine began in 1955 as an open-ended effort by the Air Force Propulsion Laboratory at Edwards Air Force Base, California, to explore the limits of liquid-fuel technology. Lacking a specific air force requirement for a large launch vehicle, Rocketdyne, at the direction of the laboratory, designed and built a test article capable of producing one million pounds of thrust. Operational testing of the new engine under air force steward-ship commenced in 1957, the same year the Soviet Union launched *Sputnik*, the world's first artificial satellite. Even by air force standards, Rocketdyne's accomplish-ment far exceeded the operational limits of existing missile propulsion systems. The three Rocketdyne engines for the air force's most ambitious rocket to date—the Atlas intercontinental ballistic missile (ICBM)—produced a maximum combined thrust of 360,000 pounds.

Two years later, in 1959, after the air force began to transition from liquid- to solid-fuel engines for ballistic missiles, NASA picked up the Rocketdyne contract and instructed company engineers to scale up the engine's thrust to 1.5 million pounds in anticipation of the requirements for a heavy launch vehicle, perhaps for human lunar landings. Like the air force, however, NASA had not yet established any specific vehicle design, leaving Rocketdyne engineers with only general guidelines for the engine configuration. NASA's only concrete objective focused on incremental improvement in propulsion technology, which drew on Rocketdyne's longstanding experience in the field.

Although indigenous rocket development organizations, such as the Jet Propulsion Laboratory at the California Institute of Technology (Caltech) and Reaction Motors, Inc., had been founded during World War II, the onset of the Cold War sparked rapid growth in the field. Aircraft manufacturers had enjoyed boom times during the war, but now they faced canceled government contracts, factory closures, and massive employee layoffs as a result of the wholesale demo-bilization of the armed forces. Diversification into new captive markets, such as rocket propulsion, helped firms like North American Aviation soften the blow of

the postwar economic downturn in the aircraft industry. Rocketdyne—originally called the Aerophysics Laboratory when North American Aviation established it in 1945 as a separate corporate R&D organization to exploit wartime German and American developments in jet and rocket propulsion technology—quickly became a leading designer and manufacturer of liquid-fuel engines for many army and air force missiles during the early years of the Cold War. Examples include, but are not limited to, the Redstone short-range ballistic missile, the Jupiter and Thor intermediate-range ballistic missiles, and the Atlas intercontinental ballistic missile.

Other firms, such as Aerojet General, an outgrowth of the Jet Propulsion Laboratory at Caltech, also entered the field and built the propulsion systems for the Titan I and II ICBMs. The Redstone, Titan, and Atlas missiles served NASA, which had been established by President Dwight Eisenhower in 1958, as dependable, first-generation launch vehicles for the agency's human spaceflight, satellite, and space probe programs. But the early Soviet lead in rocket-lifting capacity impelled the U.S. government to begin development of the Saturn family of rockets, starting in 1958. They became the launch vehicles for Apollo in the 1960s.

A U.S. Army Jupiter intermediate-range ballistic missile sits on the launch pad at Cape Canaveral, Florida. The S-3 engine that powered the Jupiter burned RP-1 fuel, just like the F-1, which owed much of its technological lineage to the S-3 and other liquid-fuel propulsion systems developed and built by Rocketdyne. *NASM*

Rocketdyne tested a prototype of the uprated F-1 engine, now under contract to NASA, in April 1961. It produced a maximum thrust of 1.6 million pounds, nearly ten times the output of the company's new H-1 engine. NASA had just acquired the H-1—capable of producing 165,000 pounds of thrust—from Rocketdyne to power the first stages of the Apollo program's two test platforms, the Saturn I and Saturn IB launch vehicles, which incorporated designs and technology from the Redstone, Jupiter, and Atlas missiles. The H-1 descended directly from the engine that powered the Jupiter and Thor intermediate-range ballistic missiles, while the F-1 incorporated much of the H-1's proven technology on a much larger scale. In only fifteen years, thrust jumped twenty-fold from the seventy-five thousand pounds of the Redstone engine—a derivative of the motor that powered the German V-2—to the million and half of the F-1. F-1 development followed a somewhat paradoxical technological trajectory from the drawing board to the factory floor. Engineers deliberately exploited technologies

"THUNDER MOUNTAIN"
ROCKETDYNE'S ENGINE TEST FACILITY

By the early 1960s, the residents of Canoga Park, California—a seemingly sleepy San Fernando Valley enclave at the base of the Santa Susana Mountains, thirty-five miles northwest of Los Angeles—had grown accustomed to the roaring sounds emanating from what the locals often called "Thunder Mountain." Here, carved deeply into the rocky terrain, the Rocketdyne Division of North American Aviation operated the Propulsion Field Laboratory, an outdoor test site for the company's expanding line of liquid-fuel rocket engines. A decade earlier, Rocketdyne had already begun testing the big booster for the air force's Navaho, a massive, supersonic intercontinental cruise missile designed to carry a nuclear warhead to a range of more than three thousand miles. The Navaho never entered the air force inventory, but the propulsion systems that followed on its heels powered many of the large military and civilian launch vehicles operated by the army, air force, and NASA after World War II.

Engine test stands fill out the rocky landscape at Rocketdyne's Santa Susana field laboratory. *University of Southern California Special Collections*

Rocketdyne carried out engine testing on a spectacular scale at Santa Susana. Noise, flames, and smoke often enveloped the site. At one extreme, engineers fired a research model thruster designed to put out a maximum thrust of twenty-five pounds for an attitude control system on board the Gemini spacecraft. At the other extreme, by contrast, engineers tested the much bigger and more powerful engines for the Redstone, Jupiter, Thor, and Atlas missiles and the Saturn family of launch vehicles. By early 1964, the three-square-mile test facility had completed two hundred thousand engine tests, many of them on half a dozen large stands—"metal-scaffold towers resembling oversize oil derricks," the *New York Times* reported at that time. By the end of the decade, the number of test stands had tripled to eighteen.

Large-scale engine testing continued into the 1970s for NASA's Space Shuttle. In 1978, Rocketdyne began testing its new main engine, rated to produce four hundred thousand pounds of thrust. The program ran for ten years, until 1988, when NASA moved shuttle main engine testing to the agency's John C. Stennis Space Center, located inland from the Mississippi coast east of New Orleans. That center had originally been built for the operational testing of Saturn V stages during the Apollo program. By then, the number of test firings at Santa Susana had dropped sharply, due in large part to the Defense Department's permanent shift away from liquid to solid propellants for long-range missiles. The Santa Susana site permanently closed shortly after the Boeing Company purchased Rocketdyne from Rockwell International (which had acquired North American in the late 1960s) in 1996.

A U.S. Army Redstone missile, essentially a larger and more powerful version of von Braun's V-2, lifts off the launch pad at the White Sands Missile Range in New Mexico. *NASM*

proven to work well on previously developed engines, but the huge rise in thrust necessitated major advances in component design, material development, and fabrication to accommodate the increased pressures, speeds, and temperature gradients of the engine components in operation.

Even massive new test stands made of reinforced concrete and steel framing had to be built to accommodate such a large engine for live firings. In its first public demonstration in August 1961, an F-1 engine captivated viewers at Edwards Air Force Base. Mounted vertically on an eleven-story stand perched atop a granite cliff, the engine blasted blinding flames—and deafening noise—down a five-hundred-foot wall dug into the mountainside.

Like the H-1, the F-1 burned RP-1 (short for "rocket propellant"), a highly refined form of kerosene similar to jet fuel. Nonvolatile and storable at room temperature, RP-1 had a long service life as the fuel of choice for rocket engines, thereby fulfilling NASA's mandate for safety and reliability. The Rocketdyne and Aerojet General engines for the Thor, Jupiter, Atlas, and Titan I missiles also used kerosene. Its known properties enabled engineers at Rocketdyne and NASA to avoid some of the hazards and technical demands of storage and handling otherwise placed on engine components that burned more exotic and less easily storable fuels, such as ultra-cold liquid hydrogen. NASA did, however, opt for a liquid hydrogen engine for the second (S-II) and third (S-IVB) stages of the Saturn V, largely because of the much higher specific impulse—essentially, higher performance per unit of fuel burned—needed to lift the Apollo spacecraft to the Moon. Rocketdyne designed and built this engine too, designated the J-2 and rated to produce two hundred thousand pounds of thrust (later increased).

Like its predecessors, the F-1 used liquid oxygen as the oxidizer. But in terms of scale, the F-1's combustion chamber covered an area nearly four times larger than that in the H-1. It consumed more than fifteen thousand gallons of RP-1 and nearly twenty-five thousand gallons of liquid oxygen per minute, or a combined total of 3.3 tons of fuel and oxidizer per second. Rocketdyne engineers also incorporated a removable, gas-cooled nozzle extension into the design to cool the engine and simplify transport—no easy task given the F-1's size and weight.

Saturn V S-IC stages under construction at NASA's Michoud Assembly Facility near New Orleans, Louisiana, are prepared for F-1 engine installation on October 1, 1968. *NASA*

Despite NASA's decision to use proven kerosene-oxygen technologies in the F-1, engineers still faced formidable obstacles during the transition from test article to production model, especially given the huge size and power of the engine. Shortcomings in materials design and performance manifested themselves as the program matured, notably during testing. Even the spectators present at the first public firing of an F-1 at Edwards Air Force Base in August 1961 could not help but feel disappointed when a fuel system malfunction forced a shutdown after just 1.5 seconds. A more detailed discussion of the evolution of some of the major components highlights the innovations that NASA and Rocketdyne engineers incorporated into the final design.

To get some sense of the complexity of their work, it is useful to think of the F-1, or any rocket engine for that matter, as a controlled explosion. Propellants enter the combustion chamber through the injector at temperatures exceeding several hundred degrees below zero—in the case of the F-1, only the oxidizer is super-cooled—and the hot exhaust exits the other end at several thousand degrees. This temperature gradient, in addition to the extreme pressures of propellant flow and combustion, put enormous strain on the engine components and materials. Engineers had to strike a delicate balance between safety and reliability—that is, working within the technological state of the art—and the innovations needed to meet the propulsion system's minimum operational requirements. As aerospace historian Roger Bilstein has noted, "Once designers got into advanced metallurgy, they got into innovation. Coupled with the factors of size and operating requirements of the F-1, there ensued a number of

The first Saturn V flight in November 1967 marked the culmination of a development process that began in the late 1950s, prior to NASA's establishment and several years before the nation committed itself to sending a human mission to the Moon. The development, production, and testing of the Saturn I and Saturn IB boosters established the requisite learning curve—covering everything from propulsion to staging to guidance and control—that engineers from NASA and a host of industrial contractors needed on their way to the Saturn V. Although NASA had intended from the outset to use the Saturn family of launch vehicles to send humans into space—a first for the new space agency—the bulk of the technical effort focused on scaling up technologies already in use on other rockets and missiles. That is not to say significant innovations did not result from this process. To the contrary, NASA and its contractors achieved some major milestones, but they all had some basis in prior technological developments.

Work on what became the Saturn I booster originated at the U.S. Army Ballistic Missile Agency in 1958. (The core of that agency—Wernher von Braun's Development Operations Division—became NASA's George C. Marshall Space Flight Center in July 1960.) Von Braun's team proposed a three-stage launch vehicle, simply called Saturn, to put the United States on par with the Soviet Union in rocket power. Eight Rocketdyne H-1 engines, each with an initial thrust rating of 165,000 pounds (later upgraded to 188,000 pounds), powered the boost stage (S-I), giving a total thrust of 1.3 million pounds, the highest thrust at that time for a single launch vehicle. To save time and money, von Braun's engineers adapted the tooling

Shown here is a completed Saturn 1 launch vehicle in the Fabrication and Assembly Engineering Division at the George C. Marshall Space Flight Center on February 1, 1961. The vehicle is in its original configuration, which included two dummy upper stages. *NASA*

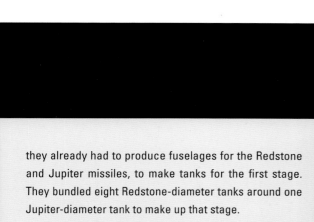

they already had to produce fuselages for the Redstone and Jupiter missiles, to make tanks for the first stage. They bundled eight Redstone-diameter tanks around one Jupiter-diameter tank to make up that stage.

Once the Saturn program became part of Apollo, the Saturn I turned into a two-stage vehicle with a liquid-hydrogen/liquid-oxygen upper stage. The S-IV (the numbering is confusing and represents the history of the program) ran off a cluster of six RL-10 engines, the world's first operational liquid-hydrogen rocket engine and the first designed to restart in flight. The Pratt and Whitney Division of the United Aircraft Corporation, a well-known maker of jet engines, had developed the RL-10 for the Atlas-Centaur rocket. The first Saturn I lifted off the launch pad at Cape Canaveral, Florida, on October 27, 1961, without the as-yet-to-be-developed second stage. Nine more flights followed up to July 1965; the fifth, in January 1964, flew the S-IV for the first time. The seventh flight, in September 1964, placed a prototype Apollo spacecraft into orbit.

The second Saturn 1B to carry an Apollo spacecraft prepares to clear the launch tower at Cape Canaveral, Florida, on August 25, 1966. Mission objectives included confirmation of predetermined launch loads, demonstration of spacecraft component separation, and verification of heat shield strength at high reentry rates. *NASA*

The Saturn IB, a scaled-up Saturn I, had a larger and more powerful second stage. NASA announced the new rocket, also developed by Marshall, on July 11, 1962. The S-IVB stage served, with modification, as the third stage of the Saturn V, "exemplifying," as aerospace historian J. D. Hunley has written, "the building-block nature of the development process." Like its predecessor, the Saturn IB relied on eight upgraded H-1 engines (first rated at 200,000 pounds and then 205,000 pounds) to power the S-IB first stage. NASA, however, jettisoned the RL-10 engine in favor of the new Rocketdyne J-2 engine for the S-IVB second stage. Like the RL-10, the J-2 ran on liquid hydrogen fuel, but its greater size and thrust provided a clear operational advantage. A single J-2 produced more than twice as much thrust—at first two hundred thousand pounds, increasing later—than all six RL-10s combined. The first Saturn 1B flew on February 26, 1966, followed by four more flights to test the launch vehicle itself and also the Apollo spacecraft—command, lunar, and service modules—designated to fly on the Saturn V. Saturn IBs launched five human missions: Apollo 7 in 1968, three Skylab space-station crew flights in 1973, and the Apollo-Soyuz Test Project mission to link up with a Soviet spacecraft in 1975.

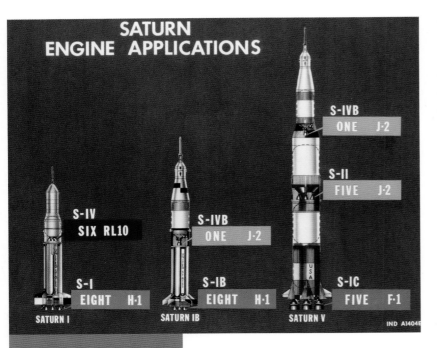

SATURN
ENGINE APPLICATIONS

S-IV
SIX RL10

S-I
EIGHT H-1

SATURN I

S-IVB
ONE J-2

S-IB
EIGHT H-1

SATURN IB

S-IVB
ONE J-2

S-II
FIVE J-2

S-IC
FIVE F-1

SATURN V

IND A1404E

This illustration shows the staging and engine configurations for the Saturn I and IB and Saturn V rockets. *NASA*

technological advances and innovations in fabrication techniques."

For example, Rocketdyne engineers tried to scale up the injector used in the H-1, but that approach failed to yield a workable solution. The F-1's injector sprayed the RP-1 and liquid oxygen through more than six thousand separate holes in a specific pattern intended to maximize efficient burning inside the cylindrical combustion chamber. Combustion instability became increasingly problematic during testing, until one engine suffered a complete meltdown at Edwards Air Force Base in June 1962. A major redesign, which included enlarging the diameters of the injector orifices, improved stability. In subsequent testing, Rocketdyne and NASA engineers deliberately destabilized combustion by detonating small explosive charges inside the engine and monitored how fast the engine recovered its stability. Given the complexity of calculating theoretically the most efficient and stable combustion process, their effort constituted a combination of art and engineering.

Another case is the turbopump, which sucked the propellants out of the tanks at a rate equivalent to draining a large swimming pool in a matter of seconds. It weighed in at more than a ton and generated more than 50,000 horsepower. Upgrades to new high-strength materials yielded problems, such as weaknesses in the welding spots in the pump manifold. In this case, Rocketdyne had used a new nickel-based alloy developed by General Electric. Engineers devised welding techniques to mitigate cracking and developed new training protocols for welders on the F-1 production line.

On April 16, 1965, ten years after the air force let the first contract for the F-1, engineers fired all five engines of the Saturn V first stage at the Marshall Space Flight Center in Huntsville, Alabama, for 6.5 seconds. In early August, engineers completed a longer test equivalent to the entire first-stage burn—2.5 minutes. Those engines specifically designated for Saturn V launches underwent longer tests lasting 10.8 minutes. NASA had specified that the F-1 should be capable of running for 24.3 minutes over the course of its lifetime, including testing prior to launch, once again highlighting the overriding importance of durability and reliability in the Apollo program. In September 1966, NASA qualified the F-1 for human missions. In all, it underwent nearly three thousand tests, almost half of them for durations exceeding the intended flight time of 2.5 minutes.

On November 9, 1967, the first flight-ready Saturn V lifted off the pad at NASA's Kennedy Space Center, Florida, on the Apollo 4 mission. All five F-1s performed flawlessly, and the flight went off without a hitch. The engines on the S-1C first stage shut

A Saturn V S-IC stage is lifted onto a test stand at the Mississippi test facility, now the John C. Stennis Spaceflight Center, on January 1, 1967. *NASA*

down in sequence, first the center engine at 135.5 seconds, then the four outboard F-1s at 150.8 seconds. At this point, when the J-2 engines in the second stage ignited, the vehicle had reached a velocity of 6,000 mph at an altitude of thirty-eight miles.

In April 1968, the next test of the Saturn V—designated Apollo 6—did not yield such a pristine result, however. Severe vibrations—called the "pogo effect"—proceeded along the length of the rocket during ascent, both during the first and the second-stage firings, while the upper stages also experienced J-2 engine failures. The F-1 engines developed natural vibrations at a frequency that matched the oscillations in the airframe, thereby creating a cumulative effect that intensified at the top of the launch vehicle, where the crew compartment sat. Engineers eliminated the possibility of a recurrence in future flights by inserting a shock absorber—in this case, helium gas—into the valve assembly connecting the liquid-oxygen lines to the engines. When the valve opened, the helium gas dampened the engine pulsations and prevented the vibrations from spreading into the remainder of the airframe. Confident in their solution, NASA Marshall engineers and managers proceeded directly to the next Saturn V launch, which carried the Apollo 8 crew—the first humans—to the Moon in December 1968. That flight, and all others,

The space shuttle *Atlantis* heads toward low Earth orbit on March 24, 1992. Note the three main engines attached to the aft section of the orbiter. They burned for 8.5 minutes during ascent, generating a combined thrust of approximately 1.2 million pounds. *NASA*

registered minimal pogo effects, and the F-1s performed as expected.

Over the course of the Apollo program, Rocketdyne built and delivered to NASA ninety-eight F-1 engines, sixty-five of which powered thirteen Saturn V vehicles—twelve for Apollo and the last for the Skylab space station. Budget cuts at NASA brought the Apollo and Skylab programs to a close in the early 1970s, thereby eliminating any prospects for continued F-1 development or use. Opting for economical and routine spaceflight, NASA switched from disposable to reusable rocketry and moved ahead with the development of the Space Shuttle. Unlike the single-use F-1 and all of its liquid-fuel predecessors, reusable rocket engines powered the Space Shuttle during launch and ascent. Also unlike the F-1, the Space Shuttle main engine extended the limits of the technological state of the art. Assembled from fully interchangeable parts, it produced less than one third the maximum thrust of the F-1, but it burned fuel—in this case, super-cooled liquid hydrogen—much more efficiently.

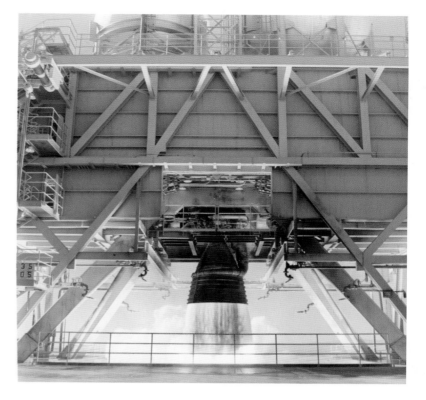

The National Air and Space Museum's F-1 engine undergoes static testing on February 16, 1965. *NASM*

Displaced by the Space Shuttle main engine, the F-1 passed into history. Since the last Apollo lunar mission in 1972, the remaining F-1 engines have served as exhibition pieces on public display in museums throughout the United States and Europe. The National Air and Space Museum acquired about half a dozen F-1 engines after the end of the Apollo program. Most are currently on loan elsewhere. The most historically significant F-1, built by Rocketdyne in 1963 and donated to the Smithsonian by the Marshall Space Flight Center in 1970, is currently on display in the Museum's Apollo to the Moon gallery. NASA tested this engine four times, for a total duration of 192 seconds.

Another F-1 engine, which the Museum acquired from NASA in 2004 and displayed at the Steven F. Udvar-Hazy Center in Virginia, has since been deaccessioned and returned to the Marshall Spaceflight Center. There it underwent refurbishment and testing in 2012, nearly sixty years after the air force issued the contract to Rocketdyne to develop the technology. Engineers at Marshall, many of whom had been born after the Apollo program ended, studied ways to repurpose the engine and its components as part of a larger project to develop new propulsion systems for deep-space exploration. In the meantime, the F-1 still remains the highest thrust single-chamber rocket engine ever flown.

—*Thomas C. Lassman*

F-1 ENGINE
SPECIFICATIONS

MANUFACTURER:
Rocketdyne Division of North American Aviation
THRUST (SEA LEVEL):
1.5 million lbs. (6.675 million N)
THRUST DURATION (SATURN V S-1C STAGE): 150 sec.
WEIGHT (DRY): 18,416 lbs. (81,951 N)
PROPELLANTS: RP-1 (kerosene) and LOX (liquid oxygen)
HEIGHT: 18.5 ft. (5.64m)
DIAMETER (BASE OF NOZZLE):
12.2 ft. (3.72m)

SATURN V LAUNCHES

Vehicle Number	Mission Name	Date	Crew	Mission Notes and Firsts
AS-501	Apollo 4	Nov. 9, 1967	none	First launch of Saturn V; entirely successful, lunar velocity reentry test of Apollo 4 CM
AS-502	Apollo 6	April 4, 1968	none	Launch problems with Pogo and with J-2 failures on S-II and S-IVB. S-IVB would not restart, reentry test of Apollo 6 CM done with spacecraft propulsion
AS-503	Apollo 8	Dec. 21, 1968	F. Borman, J. Lovell, W. Anders	First crewed Saturn V; first human mission into deep space and to the Moon; 10 lunar orbits and Christmas Eve TV broadcast; duration: 6 days
AS-504	Apollo 9	March 3, 1969	J. McDivitt, D. Scott, R. Schweikart	First crewed LM test in Earth orbit; EVA test of lunar spacesuit and first rendezvous and docking between LM and CSM; duration: 10 days
AS-505	Apollo 10	May 18, 1969	T. Stafford, J. Young, E. Cernan	Full dress rehearsal for Apollo 11 lunar landing in lunar orbit. Stafford and Cernan descend with LM to within 9 miles of Moon; duration: 8 days
AS-506	Apollo 11	July 16, 1969	N. Armstrong, M. Collins, E. Aldrin	First human lunar landing by Armstrong and Aldrin, Sea of Tranquility; first EVA on Moon and return of lunar samples; duration: 8 days
AS-507	Apollo 12	Nov. 14, 1969	C. Conrad, R. Gordon, A. Bean	Hit by lightning during launch; second lunar landing, Conrad and Bean, Sea of Storms, precision landing near Surveyor 3 spacecraft; return of samples and parts of Surveyor 3; duration: 10 days
AS-508	Apollo 13	April 11, 1970	J. Lovell, J. Swigert, F. Haise	Failure of center engine on S-II; oxygen tank explosion in SM on way to Moon; loop around Moon; emergency return with LM as lifeboat; duration: 6 days

Vehicle Number	Mission Name	Date	Crew	Mission Notes and Firsts
AS-509	Apollo 14	Jan. 31, 1971	A. Shepard, S. Roosa, E. Mitchell	3rd lunar landing, Shepard and Mitchell at Fra Mauro site intended for Apollo 13; more extensive EVAs and scientific work; duration: 9 days
AS-510	Apollo 15	July 26, 1971	D. Scott, A. Worden, J. Irwin	4th lunar landing by Scott and Irwin, Hadley Apennines; first mission to carry a lunar rover and do 3 EVAs; major scientific survey instruments in SM operated by Worden in lunar orbit; duration: 12 days
AS-511	Apollo 16	April 16, 1972	J. Young, T. Mattingly, C. Duke	5th lunar landing by Young and Duke at Descartes, only highland landing site; 2nd lunar rover and orbital science mission. Mission duration: 11 days
AS-512	Apollo 17	Dec. 7, 1972	E. Cernan, R. Evans, J. Schmitt	6th lunar landing, Cernan and Schmitt, Taurus-Littrow; Schmitt 1st scientist-astronaut and only one on Moon; 3rd lunar rover and orbital science mission. Duration: 12 days
AS-513	Skylab 1	May 14, 1973	none	1st U.S. space station; micrometeorite shield ripped off during launch; had to be rescued by first crew; occupied by three crews for 28, 56, and 84 days in 1973–74; Skylab burned up in atmosphere 1979

Abbreviations

AS = Apollo-Saturn
CM = Command Module
CSM = Command and Service Modules
EVA = Extravehicular Activity (spacewalk)
LM = Lunar Module
SM = Service Module

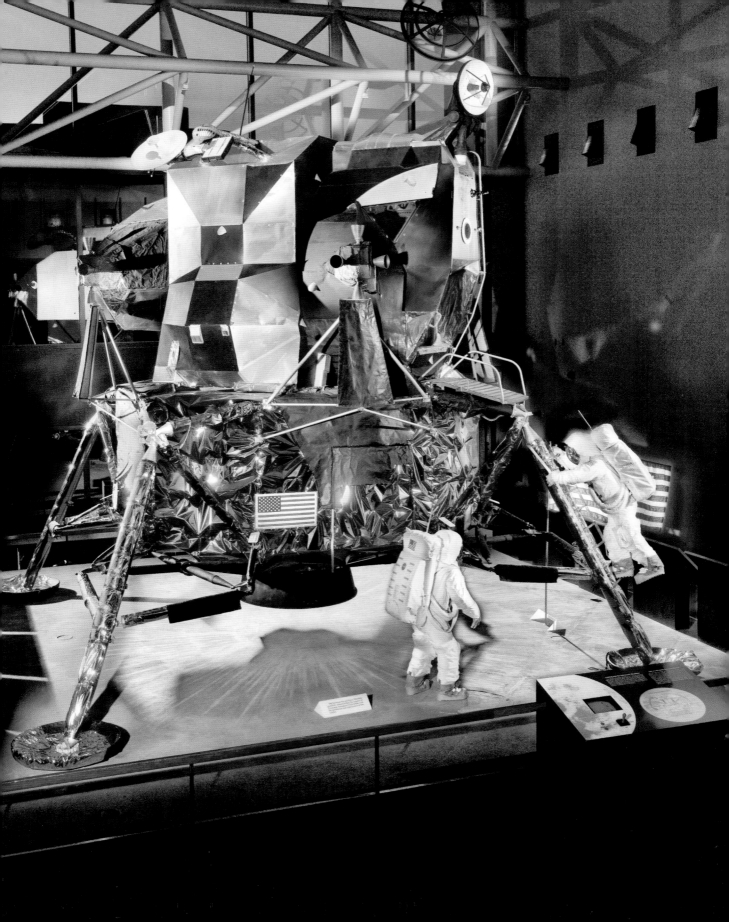

THE LUNAR MODULE (LM) astride the Moon's surface has become an iconic image of an extraordinary achievement. To commemorate that feat, accomplished in July 1969, LM-2, a vehicle once intended for an Earth-orbital test mission, has been on display at the Smithsonian since 1971. First exhibited in the Arts & Industries building near the Castle, it has become a key attraction within the new National Air and Space Museum building ever since its opening in 1976.

LM-2 was configured to resemble the Apollo 11's LM-5 *Eagle*, the spacecraft that took Neil Armstrong and Buzz Aldrin to the lunar surface. Of course, *Eagle* was unavailable. Its descent stage remains on the Sea of Tranquility, whereas its upper crew compartment (ascent stage) had crashed into the Moon days after delivering Armstrong and Aldrin to the Command Module *Columbia*, piloted by Michael Collins, for their return to Earth. Of the ten test vehicles and fifteen actual Lunar Modules planned for construction by the Grumman Aircraft Engineering Corporation in Bethpage, New York, only four have become museum pieces. LM-2, nearly complete and structurally similar to LM-5, was deemed the most historically significant artifact available in 1970 to display in the Smithsonian museum devoted to the history of flight.

The LM was one part of a complex system designed to fulfill President Kennedy's goal of getting a man to the Moon and bringing him safely back within the decade of the 1960s. It is truly remarkable that in 1961, when President Kennedy challenged

LUNAR MODULE
LM-2

/5

Congress and the American people, the United States had only fifteen minutes of human spaceflight experience. To meet Kennedy's audacious goal, NASA had to mobilize vast resources to design, build, and test countless components and support equipment. Among these were the huge Saturn V rocket, an enormous launch complex, the Apollo Command and Service Modules, and the LM.

Why a Lunar Module?

Humans who dreamed of traveling to the Moon often conceived of machines to get them there. In 1961, however, no one had yet designed anything capable of doing it. Throughout the 1950s, space enthusiasts had discussed the best method. Many imagined using what NASA planners eventually referred to as Direct Ascent: An enormous rocket would deliver to the lunar surface a spacecraft that was large enough, and had enough fuel, to take off directly for home once exploration was complete. Alternatively, Wernher von Braun, who would lead the team that created

LM-2 as LM-5 *Eagle*, the icon of the lunar landing, anchors the east end of the National Air and Space Museum and is seen by millions of visitors each year. The current exhibit is much the same as it was in Osaka, Japan, in 1970. *NASM*

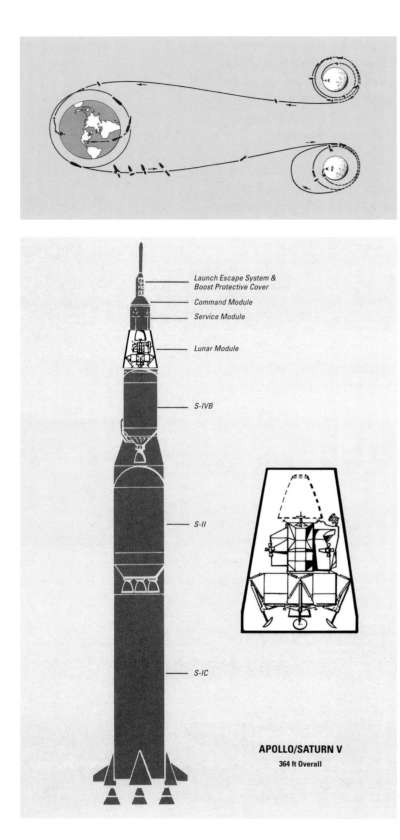

Launch Escape System &
Boost Protective Cover

Command Module

Service Module

Lunar Module

S-IVB

S-II

S-IC

APOLLO/SATURN V
364 ft Overall

the Saturn V rocket, had long envisioned sending several large components into Earth orbit, where they could be assembled for the trip to the Moon—what came to be called Earth Orbit Rendezvous (EOR). That spacecraft, like Direct Ascent, would proceed directly to the lunar surface.

After an extended, sometimes contentious deliberation in 1961–62, NASA chose a third mode: Lunar Orbit Rendezvous (LOR). A giant rocket (but not as enormous as the one needed for Direct Ascent) was to place on a lunar trajectory a modular set of smaller craft: the combined Command and Service Modules acting as a mothership, together with what was originally called a Lunar Excursion Vehicle. Once in lunar orbit, the Lunar Module, as it came to be known, would carry the astronauts to the lunar surface with an upper stage capable of launching the astronauts back into lunar orbit. Once the astronauts returned to the Command Module, they would no longer have use for the vehicle. The Service Module, which carried the main propulsion and life-support systems, would propel the Command Module and crew back to Earth and be discarded shortly before reentry. The only part of a vehicle two-thirds as high as the Washington Monument that would return to Earth was the cone-shaped Command Module.

The Grumman Corporation began designing an Apollo spacecraft in 1960 in response to a proposed NASA feasibility study, one eventually awarded to North American Aviation. Afterward, Grumman continued to pay fifty engineers to work on a vehicle for the LOR concept. This head start gave Grumman an advantage in securing the subsequent contract to build the LM, once NASA refined its requirements. The engineers at NASA and Grumman conceived of a two-part vehicle. The LM they produced was a completely self-sufficient spacecraft with life support, guidance and navigation, attitude control, communications, and all the instrumentation required for operation in lunar orbit, landing on the lunar surface, takeoff, and rendezvous with the Command Module. The LM traveled into space stowed beneath the Command and Service Modules in a special compartment atop the Saturn V called the Spacecraft LM Adapter, or SLA.

En route to the Moon, the astronauts detached the Command and Service Modules (CSM) from the four panels of the SLA, which were jettisoned. They then executed a somersault, attached the pointed end of the Command Module to the LM, and extracted it from the booster. The resulting odd-looking contraption, which in a vacuum was not subject to aerodynamics, would travel the 240,000 miles to the Moon. To orbit the Moon, the astronauts slowed the spacecraft by firing the Service Module's main engine opposite the direction of travel. Once there, they prepared to send two of the three astronauts aboard the LM to land on the Moon.

The LM had eighteen rocket engines in all: a large one for descent, another for ascent, and sixteen attitude control thrusters clustered in groups of four. The thrusters allowed the LM pilots to rotate the spacecraft into the desired position during all phases of its mission. The orbital velocity of almost 4,000 mph parallel to the surface had to be reduced so that at touchdown it was five vertical feet per second, with no lateral movement.

The LM descent stage is simple in appearance, but the engineering is complex. Because the Moon has no atmosphere, the LM had to slow down using only its engine. Rocket engineers had little experience then with throttleable engines, so it was a difficult task to design one. The descent engine also had to have a computer-controlled gimbal system to adjust the direction of thrust as the spacecraft's center of gravity changed as it consumed propellants. The four legs, with their large

OPPOSITE, TOP: Illustration of Lunar Orbit Rendezvous (LOR) depicting the spacecraft in the various phases of the mission. *NASA*

OPPOSITE, BELOW: Illustration of the Saturn V stack depicting the location of the LM for launch, just below the Command and Service Modules. *Grumman*

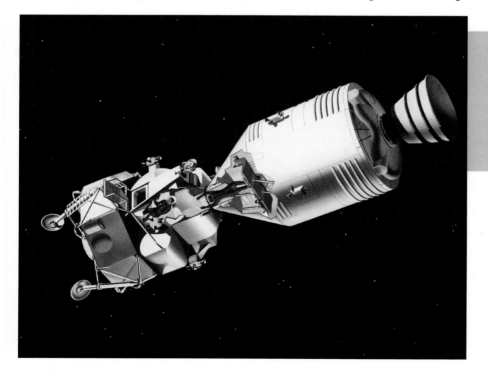

Artist concept from 1966 of the Apollo spacecraft during lunar coast phase; two of the three astronauts prepare for transfer from the Command and Service Modules to the Lunar Module. *NASA*

footpads and honeycomb aluminum structure, absorbed the final landing shock. The descent stage was also equipped with storage compartments for the experiments and equipment that the astronauts would use during their forays onto the surface.

The ascent stage housed the astronauts and the control and life-support equipment, which took two years to design. Weight was an ever-present concern. Grumman and NASA engineers considered every means of eliminating mass. They even removed the astronauts' seats, figuring that in weightlessness standing would not be overly tiring. In that position, in fact, the astronauts had better visibility and could use their legs as shock absorbers when the LM touched down. All structural material had to be as thin as possible to minimize weight.

The LM ascent stage was not only a means of transportation; it was also the living quarters for the two astronauts. Within the crew compartment, engineers provided stowage space for food and equipment as well as enough room to prepare for crucial lunar walks. The ascent-stage engine was simpler than the descent engine, as it could not fail. If it did, the two astronauts would remain on the Moon until their oxygen ran out and they perished. To minimize failures, engineers used hypergolic propellants; that is chemicals that would spontaneously combust when combined, even in the vacuum of space.

The legs and the odd clustering of shapes and surfaces on the LM made it resemble a large insect with crinkled, gold-color foil skin covering much of its lower body. Consisting of twenty-five layers of aluminized Mylar and a Kapton external sheet, these thermal blankets protected descent stage bays and the equipment from the extreme temperatures in space and the bombardment of micrometeorites. Engineers discovered that the layers of Mylar provided adequate protection, and at a fraction of the weight of the aluminum panels of the original design. To provide empty space between the layers of Mylar, technicians hand-crinkled each layer. Engineers simply taped, stapled, and screwed the blankets of Mylar to the spacecraft to keep them from falling off. Because technicians made each LM by hand, with slight modifications for each mission, they each have a unique appearance.

Eagle to the Moon

On July 20, 1969, *Eagle*, piloted by Neil Armstrong and Edwin E. "Buzz" Aldrin, was traveling at 3,800 mph in lunar orbit, 250 miles from the Sea of Tranquility. Armstrong keyed in the command to the guidance computer to begin braking. Nine minutes later, the LM was traveling at 410 mph and only ten thousand feet above the surface. NASA chose the landing site based on lunar maps and orbiter images. None had the resolution, however, to reveal whether the designated site was smooth. At just over four miles from landing, Armstrong took control of the spacecraft, so he could visually guide it to a safe landing. At eight thousand feet, he realized that the targeted spot had too many boulders. Still traveling at over twenty feet per second, Armstrong diverted the *Eagle* past a treacherous crater.

Back at Mission Control in Houston, astronaut Charlie Duke informed Armstrong that he had only sixty seconds of fuel remaining, then thirty seconds. When the probes extending from three of the pads on the descent stage touched lunar surface,

Armstrong initiated shutdown of the descent engine. The *Eagle* had landed. Six hours later, Armstrong and Aldrin stepped out onto the dusty lunar landscape, realizing one of humankind's great dreams.

The Lunar Modules

Nine years of intense work at Grumman had culminated in astronauts walking on the Moon. In June 1967, Grumman had delivered LM-1 to NASA, but it was not until January 1968 that it made its unmanned flight as the Apollo 5 mission. LM-1 circled the Earth five times while technicians remotely tested crew-compartment integrity, the attitude thrusters, throttling of the descent engine, stage separation, and ignition of the ascent engine. LM-1 had no legs, for it was not to land anywhere. Its short, successful life ended in a fireball over Panama when it reentered the atmosphere. LM-2 (now on display at NASM) was the backup, designed and constructed to the same specifications. Due to the success of the LM-1 flight, NASA managers decided not to launch LM-2. Nevertheless, they asked Grumman to finish the craft and deliver it to Houston for testing.

NASA used LM-3 (*Spider*) and LM-4 (*Snoopy*) as crewed test vehicles. LM-3 flew into space on Apollo 9 in March 1969 for an Earth-orbit rehearsal of the lunar landing. Over a ten-day period, the astronauts put the combined CSM/LM through all of the

LM-3 *Spider* orbits Earth during the Apollo 9 mission in 1969. For ten days, NASA put the LM and CSM through the phases of undocking and redocking as they would in lunar orbit. *NASA*

phases of a lunar mission. All engines were tested, and the astronauts practiced docking and re-docking of the LM and Command Module. Two months later, Apollo 10's crew took LM-4, *Snoopy*, to the Moon. It was not fully equipped to land on the Moon, but the astronauts flew it within nine miles of the lunar surface, where they helped scout out landing sites on the Sea of Tranquility. The descent stage was jettisoned and eventually crashed onto the lunar surface. After successfully docking with the Command Module, the astronauts released the ascent stage into a solar orbit, where it remains today.

LM-6 (*Intrepid*) and LM-8 (*Antares*) both successfully landed on the Moon (Apollo 12 and Apollo 14). Subsequent, so-called "J Class" mission LMs on Apollo 15, 16, and 17 (LM-10, -11, and -12) not only successfully landed on the Moon but also carried with them the Lunar Roving Vehicles, which greatly expanded the area that could be explored and the quality of scientific information collected. LM-7 (*Aquarius*), aboard Apollo 13, famously provided the astronauts with an emergency refuge after an oxygen tank in the Service Module exploded while en route to the Moon. The three astronauts survived, huddled in *Aquarius* for the return to Earth, stretching oxygen and water supplies meant for two men for two days to provide for three astronauts for four days, thereby illustrating the flexibility of the LM's design and construction.

Why Does LM-2 Look Like LM-5?

Since its completion in 1968, LM-2 has had a storied career. Months before the Apollo 11 launch, NASA remained concerned that the LM could fail during a rough lunar landing. LM-2, with added landing gear and fuel tanks filled with inert fluid, went through rigorous testing at Houston's vibration and acoustic testing facility.

TV COVERAGE OF APOLLO

During the Apollo program, the United States had other vital concerns, including the struggle for civil rights; the effort to combat poverty; urban unrest; the assassinations of President John Kennedy, his brother Robert, and Martin Luther King Jr.; and the increasingly unpopular military conflict in Vietnam. Like, and amid, these national and international events, the successes and travails of the Mercury, Gemini, and Apollo programs were broadcast to the nation via television.

Among the reporters on the air at the time, none became more closely associated with the Apollo lunar landings than CBS news anchor Walter Cronkite. Cronkite's television coverage of the Apollo lunar landings became, for many Americans, the primary way that they witnessed this history. NASA, aware that the American people needed to be a part of the almost daily unfolding of historic events, took special care—through briefings and special press materials—to introduce journalists to the technical details prior to the missions. NASA assembled a cadre of specialists who were always available for interviews.

Cronkite had a keen and infectious interest in the space program. He reported from the back of a station wagon parked at Cape Canaveral on the launch of Alan Shepard aboard the first Mercury flight. Throughout Apollo, he engaged with NASA staff and engineers to understand and convey the importance and intricacies of the program. Cronkite later donated his archives to the University of Texas at Austin, but NASM has in its collection the LM model that he used on the air during CBS News broadcasts. Cronkite used many models, some very large and complex, to help explain spaceflight to his viewers. The small LM model, most likely commercially produced by Topping Models, Inc., carries the emblems of Grumman and NASA on its circular base. Cronkite held this model during his broadcasts, which often included elaborate animation of the lunar landings provided by NASA.

In 2006, NASA awarded Cronkite with the Ambassador of Exploration Award, making him the only recipient of the honor other than astronauts and NASA employees. Upon Cronkite's death in 2009, Neil Armstrong released the statement that "Walter Cronkite seemed to enjoy the highest ratings. He had a passion for human space exploration, an enthusiasm that was contagious, and the trust of his audience. He will be missed."

Walter Cronkite used a commercially produced industry model during CBS News television broadcasts. This model is in the collection of the National Air and Space Museum. It is one of the few pieces of material culture (other than the recordings of the broadcasts themselves) that illustrate how most Americans watched the Apollo program. *NASA and NASM*

Technicians dropped LM-2 from various heights and angles to study the shock on the landing gear and the electrical wiring. Engineers deemed the tests completely successful. With testing complete, LM-2 would begin its life as a public exhibit.

Following the success of LM-1, the Smithsonian had already contacted NASA to lay the groundwork for the transfer of LM-2 to the Smithsonian for exhibition. In September 1969, the Smithsonian contacted NASA again to request transfer.

LM-2, recognizable by the triangular window, during drop tests in Houston prior to the Apollo 11 launch. Engineers dropped the LM, stripped of its Kapton blankets and protective aluminum panels, in a controlled environment to examine the spacecraft's durability under the stresses of landing. *Grumman*

Unidentified dignitaries pose in front of the Lunar Module at Expo '70 in Osaka, Japan. LM-2, configured as *Eagle*, was the centerpiece of the Sea of Tranquility exhibit. *NARA*

However, the United States Information Agency (USIA) had other plans for NASA's Apollo artifacts. Expo '70, the world's fair scheduled for Osaka, Japan, from March to September 1970, had the theme "Progress and Harmony for Mankind," and U.S. officials wanted to ensure that a large number of Apollo artifacts were in the United States pavilion.

President Eisenhower created the USIA in 1953 to promote the ideals of the United States abroad and to counter perceived misinformation from the Soviet Union. Congress kept a tight rein on the agency financially and legislatively, preventing the organization from disseminating information within the United States. In the 1960s, American space accomplishments became a major focus for USIA efforts. At Expo '67 in Montreal, Canada, exhibits included an LM mockup on a simulated Moon surface inside the U.S. pavilion, a massive geodetic dome by Buckminster Fuller.

For the display in Japan, USIA staff believed that it was important to exhibit authentic American artifacts. With American accounts of the Vietnam conflict increasingly challenged, the USIA wanted an exhibit whose accuracy could not be impugned. There would be no models or mockups, only spacecraft that had flown in space or could fly in space. The largest space exhibit was the simulated Apollo 11 landing site, and the display required an actual LM, one that could convincingly stand for the LM-5 *Eagle*. Grumman engineers mated the descent stage of LM-2, wrapped in Kapton, and with legs added, to the ascent stage of Lunar Module Test Article 8 (LTA-8). Constructed as a thermal vacuum test article used to test environmental control systems, LTA-8 was the first man-rated LM, and therefore it was as authentic as anything available for display. NASM, hoping to eventually obtain an LM to exhibit, supported the efforts of USIA in Osaka, lending other Apollo artifacts for the exhibit. The USIA exhibited a lunar rock, spacesuits, a massive F-1 engine, the Apollo 8 Command Module, and other artifacts representing Apollo's successes. More than 18 million visitors appreciatively experienced the LM in Japan, reinforcing the value of replicating *Eagle*'s appearance.

LM-2 at the Smithsonian

The Apollo program was making history with each mission, and the Smithsonian worked to ensure it would be represented within its walls. In addition to a 1967 agreement between the Smithsonian and NASA stating that NASA would transfer historic objects no longer required for its operations, NASM had taken as one of its missions to engage the public in space exploration. The Air and Space Museum organized a press event and television screening of the Apollo 11 landing on July 20, 1969. Staff vacated their offices for the news crews of NBC and CBS so that the networks could telecast the Moon landing in the Arts & Industries building (where many of NASM's space artifacts were then displayed). Throughout that warm July day and night, more than 17,000 visitors strolled under the Wright Brothers' 1903 *Flyer* and past John Glenn's *Friendship 7* Mercury capsule. At 4:17 p.m. Eastern time, they broke into cheers as the *Eagle* touched down on the Sea of Tranquility. The Smithsonian had secured a key role presenting aspects of the American space program to the public.

LM-2 on display in the two-story rotunda of the Smithsonian Arts & Industries building. This image illustrates the blast marks that technicians believed the descent would make on the lunar soil. Observations following the actual landing of *Eagle* showed that the descent engine did not produce these marks on the Moon. *NASM*

In November 1970, at the end of Osaka's Expo, the LM returned to Houston, where the Grumman technicians replaced the LTA-8 ascent stage with the one from LM-2. They also made additional modifications to the recombined LM-2, continuing the transformation of its appearance into *Eagle* for Smithsonian visitors. In April 1971, the U.S. Air Force flew the refurbished LM-2 aboard a Super Guppy transport plane to Andrews Air Force Base in Maryland for delivery. After arrival at Arts & Industries, NASA, Grumman, and Smithsonian staff spent an additional two months preparing it for display.

On June 7, 1971, NASA delivered *Columbia*, the Command Module from Apollo 11, to Arts & Industries, and in July, the Smithsonian presented it and LM-2 to the public in the two-story Rotunda. The Smithsonian invited the public to view *Columbia* with the char marks from the fiery reentry as well as an LM looking much like *Eagle*. Along with the craft, the Museum exhibited lunar tools and other artifacts as well as

Unidentified Smithsonian Institution employees and contractors disassemble LM-2 for transport to the new National Air and Space Museum. *NASM*

the actual spacesuits worn by Neil Armstrong, Buzz Aldrin, and Command Module pilot Michael Collins. Then, as now, the Smithsonian identified the LM as LM-2, not *Eagle*, but the public was encouraged to view it as closely resembling the craft that successfully landed men on the Moon.

The preceding April, Smithsonian Secretary Dillon Ripley had appointed Michael Collins director of NASM. Less than a year later, crews broke ground for the new

LM SKIN REPLACEMENT

The architects and designers of the NASM building did not take into account that the Kapton coverings on the LM's lower stage tend to degrade quickly with exposure to high light levels, which bathe the east end of the building. As early as 1977, a year after installation in the new building, NASM recognized the need to periodically replace damaged coverings.

For spaceflight, Grumman engineers covered the exposed surfaces of the LM Descent Stage with specially constructed blankets composed of very thin aluminized Mylar, with an external layer of aluminized Kapton film of varying thickness depending on the anticipated exposure to solar heating and engine exhaust. The DuPont company manufactures Mylar, a strong, lightweight polyester film. For the LM blankets, DuPont laminated the Mylar with aluminum for added heat reflection. Kapton, also manufactured by DuPont, is a polyimide film capable of withstanding temperature fluctuations from almost −500°F to 750°F above. Due to weight restrictions, the material was kept to the minimum feasible.

LM-2, configured in the Museum as *Eagle*, was not subjected to the actual hazards of space, so technicians attached either a layer of Mylar with a layer of Kapton over it or simply a single layer of Kapton for visual effect. In 1984, the first major Kapton replacement took place. Technicians from Grumman came to NASM to replace the skin on the east side of the LM with Kapton-covered plastic sheeting. Fortunately, Kapton on the west side was not subjected to direct sunlight and had remained in relatively good condition. Grumman added a layer of two mil orange Kapton to the footpads over the existing, thinner film. The darker-colored Kapton on the struts was so deteriorated that it had to be removed and replaced.

In 2009, for the fortieth anniversary of Apollo 11, NASM engaged the expertise of Paul Fjeld, who had been an official NASA artist during the Apollo-Soyuz program and an expert on the LM appearance. Rather than subjecting the LM structure to numerous removals and attachments, Fjeld proposed to fabricate four lightweight aluminum lattices with hooks that could be simply hung without any drilling into the descent stage bulkheads on the east and west sides. The blankets could then be replicated by attaching new slightly oversized Kapton sheets to the lattice, which would serve as a much more visually accurate substitute for the layered Mylar underneath. The replacement team employed various thicknesses of aluminized Kapton foil for recovering the LM, which included the landing struts and footpads.

NASM Apollo curator Allan Needell removes light-damaged Kapton from the footpad during the Kapton replacement project prior to the fortieth anniversary of Apollo 11 in 2009. *NASM*

The Lunar Landing Research Vehicle (LLRV) in flight at NASA Dryden Flight Research Center in 1964. The LLRV underwent rigorous testing at Dryden prior to transfer to the NASA Manned Space Center in Houston for astronaut training. *NASA*

The two-person LM required a unique set of piloting skills. Its engines could be controlled through twin hand controllers that sent commands through the Primary Navigation and Guidance System (PNGS), or in an emergency, through its backup, the Abort Guidance System (AGS). The Commander and the Lunar Module Pilot, who primarily monitored systems during landing missions, had an identical set of control sticks. On the right of each astronaut was an Attitude Control Assembly (ACA) for commanding the rotation (pitch, yaw, and roll) of the craft, and on the left was the Thrust/Translation Controller Assembly (TTCA) for changing horizontal and vertical velocity. During the Apollo 9 and 10 missions, astronauts practiced maneuvering, staging, and docking the LM to the Command Module, but it was not possible to practice a lunar landing while weightless in Earth or in lunar orbit.

Aerosystems designed the Lunar Landing Research Vehicle (LLRV) for testing, studying, and simulating LM approach and landings. Its first flight was on October 30, 1964, at NASA Dryden Flight Research Center on the grounds of Edwards Air Force Base in California. The aluminum-trussed LLRV, affectionately called the "Flying Bedstead," lifted a single astronaut for roughly ten minutes of simulated LM flight. It had a jet engine mounted on a vertical gimbal, which lifted the LLRV to the desired altitude. Once at altitude, in order to simulate behavior when subject to lunar gravity, the engine was throttled down so the craft would behave as if it were one-sixth of its actual weight. At this stage of a training mission, using replicas of the controls on the Lunar Module, the operator engaged two large hydrogen peroxide thrusters to control vertical speed and sixteen smaller thrusters for pitch, roll, and yaw.

While there were almost two hundred successful flights of the LLRV, it was a test vehicle and very difficult to fly. On May 6, 1968, Neil Armstrong was piloting the LLRV when, at an altitude of about two hundred feet, the craft began to pitch forward dramatically. Armstrong was not able to correct the pitch using the thrusters and was forced to activate the LLRV's ejection mechanism. The craft crashed and burned. Armstrong parachuted safely to the ground, suffering no injuries other than a bite to his tongue.

Through the entire Apollo program, all mission commanders and LM pilots flew the LLRV or the Lunar Landing Training Vehicle (LLTV), and there were two additional crashes. Neil Armstrong, accident aside, later credited his safe landing of *Eagle* on the Moon in large part to his more than thirty practice flights on the LLRV.

Michael Collins, Neil Armstrong, and Buzz Aldrin seated in front of LM-2 during a press briefing at the tenth anniversary of Apollo 11 in 1979. *NASM*

museum building. In 1975, artifacts began moving over for a July 1976 opening. NASM staff planned to exhibit the LM in the same way it was exhibited at the Arts & Industries building (much as it had been displayed in Osaka): as *Eagle* on a raised base with faux lunar dust and two suited astronaut mannequins. In the new NASM, LM-2 would not be exhibited with the Command Module but instead would be exhibited underneath engineering "back-up" versions of the Ranger, Lunar Orbiter, and Surveyor robotic spacecraft that had photographed the Moon and tested the suitability of the surface for landing.

President Gerald Ford presided over the ribbon cutting at the grand opening of the new National Air and Space Museum on July 1, 1976, and thousands of visitors swarmed in. NASM has remained one of the most visited museums in the world, and the LM has remained an important anchor and photo opportunity at the east end of the building. In July 1979, NASM hosted a huge event for the tenth anniversary of the Apollo 11 Moon landing. Former director Michael Collins posed with his partners on that lunar voyage, Buzz Aldrin and Neil Armstrong, in front of the LM. This tradition continued through all the major Apollo 11 anniversaries.

NASA and Grumman created LM-2 as a test vehicle and used it for important tests in the months prior to the success of Apollo 11. For more than forty years, LM-2 has been on exhibit as a surrogate for *Eagle*. Beginning in Osaka, then at the Arts & Industries building, and finally at the National Air and Space Museum, LM-2 has been seen, photographed, and experienced by almost 200 million visitors. Its history is significant both as an example of successful spaceflight technology and as a cultural icon. The United States wanted the world to know that the astronauts went to the Moon for all humankind, and the Smithsonian Institution will preserve and display LM-2 so people can understand the pivotal period in human history when two men first stepped onto another world.

—*Hunter Hollins and Allan Needell*

LM-2
SPECIFICATIONS

MANUFACTURER: Grumman Aircraft Engineering Corporation, Bethpage, New York
HEIGHT: 22 ft. 11 in. (6.98m)
WIDTH (DESCENT STAGE): 14 ft. 1 in. (4.29m)
WIDTH (FOOTPAD TO FOOTPAD): 31 ft. (9.45m)
WEIGHT (UNFUELED): 8,500 lbs. (3,855 kg)
WIDTH (CREW COMPARTMENT): 7 ft. 8 in. (2.34m)
LENGTH (CREW COMPARTMENT): 3 ft. 6 in. (1.1m)
DIAMETER (CREW COMPARTMENT): 8 ft. 4 in. (2.34m)
DESCENT ENGINE THRUST: 10,125 lbf (45,040 N) throttleable
ASCENT ENGINE THRUST: 3,500 lbf (16,000 N)

THE APOLLO A7-L SPACESUIT was the first machine to grant a human the freedom to explore another world outside a spacecraft. Neil Armstrong's custom-made version of it began as an engineering concept, became an anonymous assembly of complex parts, had its most famous life while being worn on the Moon, came back to rigorous examinations to test its performance and inform the next generation of spacesuit designers, and in the process became an international icon of the triumph of Project Apollo. Its most famous image was the one broadcast from the lunar surface on July 20, 1969. Even to this day, as we approach the fiftieth anniversary of the landings, this artifact continues to be a resource for the chemical, scientific, and historical study of human spaceflight.

Preparations for the first Moon walk had been tightly choreographed. After three hours instead of two, Neil Armstrong began to position himself to exit the Lunar Module (LM). It had taken longer than their numerous dress rehearsals at home because both Armstrong and his co-pilot, Edwin E. Aldrin, were acutely aware of their situation. They felt like two football fullbacks in pads moving around in a pup tent. The LM's skin was thin and the quarters close. As lunar gravitation was only one-sixth of that of the Earth, the two men moved self-consciously while carrying a mass that would have weighed 189 pounds (eighty-six kilograms) back home.

They were especially aware of the Personal Life Support System (PLSS) backpack that extended nearly a foot from their backs and stretched from their hips

OPPOSITE: Neil Armstrong's suit fully assembled, as it was on display at the Air and Space Museum for nearly thirty years. *NASM*

NEIL ARMSTRONG'S
A7-L SPACESUIT /6

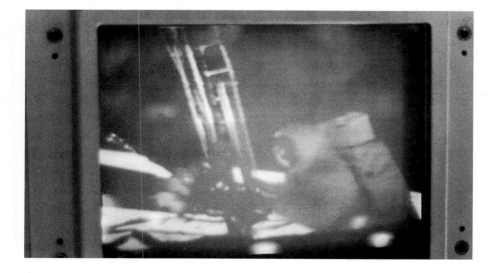

The television image of Neil Armstrong making his first small step onto the surface of the Moon. *NASA*

Armstrong, Buzz Aldrin, and Michael Collins walk toward the launch pad after suiting up for the Apollo 11 mission. *NASA*

Armstrong, Buzz Aldrin, and Michael Collins walk toward the launch pad after suiting up for the Apollo 11 mission. *NASA*

to their shoulders. It would be too easy to bump into switches and controls in the cockpit. As commander, Armstrong's left-hand position in the LM (as viewed from inside) placed him in the best position to enter the square hatch that opened to the right. There was not enough room for the two to trade places easily and in any case their bosses at NASA had long ago confirmed Armstrong's wish to be first to step foot on the Moon. He got down on his hands and knees, backed carefully onto the platform outside the hatch, and descended backward down the ladder.

For millions who watched the Apollo 11 landing and Neil Armstrong's first steps, the image of his mostly white suit will remain frozen in time. The high contrast of the black-and-white television transmission darkened the shadows underneath the LM. The folds of his spacesuit and ghosting in the image exaggerated the awkward-ness of Moon walking. As a consequence, the public's most immediate memory of the Apollo suit would not be the decades of suit design that preceded the Moon walk but those few short, waddling hours on television. Armstrong made careful and deliberate movements while exiting the LM and while moving about on the surface that contrasted strongly to the confident swagger that had characterized his and his crewmates' suited walk to the launch pad four and a half days before.

Under Earth atmosphere and gravity, the scene of the Apollo 11 astronauts walking out to the launch vehicle looked little different from what the average person might expect. The immediate public reaction to seeing the broadcast from the Moon was to characterize the spacesuits as bulky and awkward. The size of the suit was an easy target to blame for what people saw on their television screens. Prime contractor ILC Industries and NASA's Manned Spacecraft Center responded almost immediately to the "bulky" comments with detailed explanations of the relative pliability and flexibility of the suits when fully pressurized in a vacuum. This was the first time that the two organizations joined together to popularize the story of the twenty-one layers of the suit. Although it was a simplification of suit construction, the story explained why a spacesuit that appeared pliable and comfortable here on Earth could become rigid and difficult to move when worn on the lunar surface.

Armstrong and Aldrin's two hours and forty minutes on the Moon was but one chapter of the life of the A7-L (A = Apollo, 7 = 7th version, L = ILC Industries). The suit began its epic journey soon after President Kennedy announced his lunar goal in 1961. Sending humans to the Moon required advances in significant technologies, including revising existing concepts of how suits would work and how astronauts would operate in space. With only eight or nine years to complete the mission, technicians and engineers would have to compress into a few years technological developments equivalent to the achievements of the previous half century. The race to produce the Apollo spacesuit began with NASA's announcement of the initial competition in 1962. Nine companies and their partners produced prototypes for what might become the Apollo suit. Their competition would become fraught with as much bureaucratic drama as the flight itself. NASA, corporations, and the armed services had to work out very quickly how best to dress an astronaut on the surface of the Moon.

The first challenge was to determine how many functions it would fulfill. Mercury program suits were rescue suits that would have filled with oxygen if capsule life support had failed. Gemini suits filled that role during launch and entry, but also doubled as a personal spacecraft during Extravehicular Activity (EVA). The fourteen-day Gemini VII mission, in which the astronauts wore special soft suits designed to be removed in flight, made it clear that spacesuits could not function as work clothes for long periods. Therefore, the Apollo suit would have to be both a launch-and-entry rescue suit and a true spacesuit, yet easy to put on and take off so that astronauts could change their clothes when needed. The original NASA plan was to create two types, the first for Earth and lunar orbital missions and the second for walking on the Moon. This division of duties in the spacesuit mimicked North American Aviation's distinction between Block I and Block II Apollo spacecraft, which had supported their proposal to accelerate the orbital testing of the Apollo hardware. As a result, the initial suit designs for Block I suits were built on the experiences of the Gemini program, assuming that operations outside of the spacecraft would be conducted only while in Earth or lunar orbit.

Of all the prototypes created for the Block II suit, NASA engineers, technicians, astronauts, and spacecraft designers tended to prefer one from a small company in Dover, Delaware. The Special Products Division (SPD) of the International Latex Company (ILC) had grown out of the need to create display stands for the company's plastic products division, including women's undergarments and rubber gloves. ILC SPD was not entirely unknown to the pressure suit community. In late 1950s, it proposed a suit for the U.S. Air Force's high-altitude program. That design echoed a design that the Arrowhead Company had proposed to the navy in a challenge to the B.F. Goodrich Company's own Mark IV. The Arrowhead Mark IV pressure suit solved the problem of joint mobility through a series of "tomato worm" segments at

The A1-C spacesuit that Thomas Stafford used to train for the Apollo program before the Apollo 204 fire. *NASM*

The ILC SPD pressure suit submitted to the U.S. Air Force during the 1950s. *ILC Dover*

A microscopic view of the dark spots on the right glove gauntlet shows the snags, repairs, and lunar dust locked underneath. *NASM*

In 2006, the Air and Space Museum removed Neil Armstrong's spacesuit and its components from display. Curators and conservators had long been aware that the spacesuit collection was beginning to show signs of deterioration that were not only the result of long-term exhibit. The brass zippers that surrounded the rubber gaskets that held air inside the suits often had begun to corrode and even stick open or shut. The gaskets themselves had deteriorated to the point of crumbling, even with careful movement.

Museum technicians could hear crunching sounds for the first time as the layers of the Thermal Micrometeoroid Garments were beginning the wear against one another. Spacesuits in long-term cold storage had begun to flatten because of gravity. Occasionally, new stains would appear, a sign that the polyvinyl chloride (PVC) tubing inside was breaking down and leaking hydrochloric acid. The Armstrong Extravehicular Mobility Unit would become the highest priority in designing the best possible storage conditions for Apollo spacesuits, in a race to preserve this national treasure.

Conservators had inspected it twice in the decade before 2006 and had concluded that it was in remarkably good condition when compared to others that had moved from exhibition to exhibition more frequently. Its many layers had retained some degree of flexibility. After Armstrong's EMU returned to Houston in 1969, NASA dry-cleaned it after advice from Smithsonian conservators. Forty years hence, that chemical process has caused no ill effects. But the technicians undertook other actions that had impacts decades later.

When Neil Armstrong died in 2012, the Air and Space Museum decided to pay tribute to his life by displaying his extravehicular helmet and gloves at the Stephen F. Udvar-Hazy Center. Visitors and staff immediately noted the appearance of spots on the gauntlets of the gloves.

the major joints. These joints localized air displacement in the suit; as one side of the joint expanded, the other side contracted. Navy pilots disliked the Arrowhead design because the tomato worm segments were made of hard plastic that was uncomfortable after hours in the cockpit. The air force rejected ILC's suit design but accepted their bid for the X-15 helmet.

In spite of NASA's initial enthusiasm, none of the nine prototypes in the 1962 competition were able to meet the specifications for the Apollo program. They failed in four areas and brought a flaw in NASA's planning to light. The shoulders were too wide, and thus the suit did not fit the Command Module (CM) couches. The joints and gloves lacked adequate mobility to allow the astronauts to do meaningful work on the lunar surface. Helmets lacked downward visibility, so the astronauts would not be able to see their feet, which was a requisite for

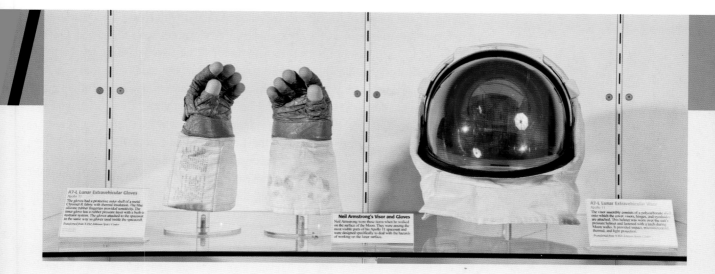

Within the image, labels read:

A7-L Lunar Extravehicular Gloves
Apollo 11
The gloves had a protective outer shell of a metal Chromel-R fabric with thermal insulation. The blue silicone rubber fingertips provided sensitivity. The inner glove has a rubber pressure layer with a built-in restraint system. The gloves attached to the spacesuit in the same way as gloves used inside the spacecraft. *Transferred from NASA, Johnson Space Center*

Neil Armstrong's Visor and Gloves
Neil Armstrong wore these items when he walked on the surface of the Moon. They were among the most visible parts of his Apollo 11 spacesuit and were designed specifically to deal with the hazards of working on the lunar surface.

A7-L Lunar Extravehicular Visor
Apollo 11
The visor assembly consists of a polycarbonate shell onto which the cover, visors, hinges, and eyeshades are attached. This helmet was worn over the suit's pressure helmet and fastened with a latch during Moon walks. It provided impact, micrometeoroid, thermal, and light protection. *Transferred from NASA, Johnson Space Center*

(Like other astronauts slated for Moon walks, he had two sets of gloves, one for outside and one for inside the cockpit). The gauntlets are an extension of the thermal micrometeoroid garment that covers the wrist connects, keeping the hardware safe and the aluminum from overheating in the unfiltered lunar sun. Previous conservation reports have remarked on the presence of some added material in several locations along the gauntlet.

By examining photos in these previous conservation reports, the Museum was able to determine that there had been a darkening in the unknown coating that NASA technicians had applied to snags in the Beta cloth cover layer. Close examination of the gloves under a high-resolution microscope revealed new information regarding their history. The darkening was sealant placed over small snags in the right gauntlet only. There was also lunar dust trapped under the repair coating. Now the Museum has new questions: Why did Neil Armstrong snag only the right gauntlet and not the left? Why did none of the other Moon walkers experience similar noncritical damage to their spacesuits? Who made the repairs and why? Perhaps future research will reveal the answers.

Armstrong's EV Visor and gloves on display at the Hazy Center in 2012, in commemoration of the astronaut's death. *NASM*

moving around on the Moon. Each prototype demonstrated unacceptable levels of air-pressure loss and restraint failures. All of the prototypes had complied with NASA's initial idea of using a separate cover garment, but through testing, the agency quickly discovered that the addition of an overcoat would make dressing and movement in the spacecraft uncomfortable.

NASA awarded the Apollo contract for the Block II Apollo spacesuit to ILC SPD as a subcontractor to Hamilton Standard. The agency's rationale was that the larger and far more experienced government contractor would best be able to handle the reporting requirements of systems engineering. Their hope was that the two companies would pool their strengths to meet agency requirements. The pairing went poorly. The two companies had numerous disputes about design responsibility, and the partnership dissolved in 1965.

The left glove dip form from which Armstrong's left gloves were made. Spacesuit gloves have to fit snugly to allow the astronaut to perform useful work. *NASM*

The falling out between ILC SPD and Hamilton Standard left NASA in a desperate position. In 1964, the agency issued a sole-source contract with the David Clark Company (DCC) to build the Block I Apollo spacesuit that would serve for Earth-orbiting missions only. It would buy NASA time to restart the bidding process for a true Moon-walking unit without having an impact on the program schedule. The Apollo launch pad fire on January 27, 1967, which killed astronauts Virgil "Gus" Grissom, Ed White, and Roger Chaffee, dealt NASA's plans a severe setback. Safety concerns forced a redesign of the crew cabins of the Command and Lunar Modules and an increase in fire resistance of everything that went into them, including the suit.

The timing of the fire and the spacecraft redesign added urgency to the bidding process on the Moon-walking spacesuit. Only three companies bid on the Block II suit: the David Clark Company, ILC, and Hamilton Standard in conjunction with B.F. Goodrich. Once again, the design merits of the ILC proposal exceeded those of its competitors. This time, NASA granted ILC the primary contract, with Hamilton Standard taking on the Portable Life Support System (PLSS) backpack and systems integration as a subcontractor to ILC. There were fewer than two years between the final contractor selection and the first flight of ILC's suit onboard Apollo 7, the first piloted mission after the fire, in October 1968. And only nine months later, the suit would be supporting life on the surface of the Moon. Everything had to work the first time.

Neil Armstrong's spacesuit began its material life under the code name "Sirius." In a May 1968 memorandum, the chief of the Apollo Support Branch at NASA assigned code names to each of the members of the Apollo astronaut pool. This enabled NASA to begin ordering suits prior to the official announcement of the crews and backups for the Apollo missions. The secret was shared by the heads of the astronaut office, the crew systems division, and the spacesuit chief contractor. At NASA's command, ILC commenced production of three spacesuits to match Neil Armstrong's personal dimensions. The interior and exterior layers, made separately and then stitched together, had separate serial numbers. Armstrong's had serial numbers 056 on the interior layer and 063 on the outer garment.

NASA's planning for training, missions, and contingencies to the Moon dictated that there would be training, flight, and backup suits for each member of the primary crew, and training and flight suits for the backup crew. For each of the main crew, three identical suits would have to be custom-made, based on a litany of body measurements that would challenge even the finest tailors. After construction, each astronaut received extensive training on how to personalize the fit for comfort and mobility. Gloves were molded from a rubber dip form that ILC had cast of each of the astronaut's hands. These custom build suits were expensive—costing approximately $100,000 each—for a total of a half-million dollars per seat in the Command Module. Later in the program, changes and modifications in the suit raised the price per suit to $250,000. NASA partially compensated for the rise in cost by reusing existing training suits. They might have cost much more, however. The price reflected a directive in the final contract that addressed both cost and urgency: Apollo spacesuits had to be built only with existing materials. There was neither time nor money to develop and test new textiles.

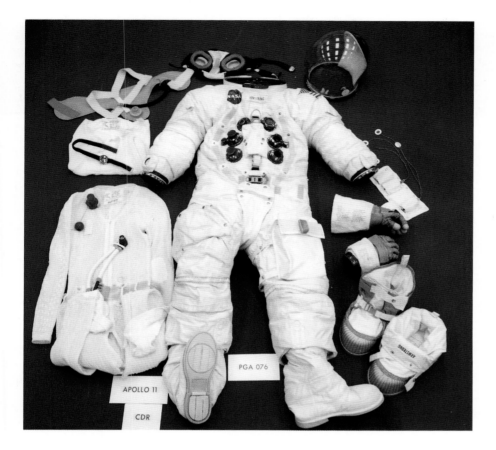

Armstrong's Apollo 11 kit, including all the layers of equipment that he wore for his walk on the Moon. *NASA*

This mandate on using existing materials and the new, post-fire flammability requirements forced ILC to combine supplies traditionally used in pilots' flight suits with other textiles, in combinations they had never considered before. Chemical companies, such as DuPont, had manufactured post-World War II textiles with the aim of carving out a place for themselves in the new, suburban economy. Dacron, Mylar, Kapton, and Beta Marcasite might have found a role in the upholstery and home-design industries alone, but together in repeated layers they could combine into a protective garment that would repel radiation from the Sun, as well as changes in temperature and penetration from a micrometeoroid traveling at 18,000 mph.

At the time of the Apollo program, the Corning Company was marketing Fiberglas curtains to homeowners through advertisements touting its resistance to fire. Once those fibers were coated in Teflon, the resulting betacloth made a fire-resistant, reflective fabric that could be cut and configured into a covering that would protect astronauts in the spacecraft and on the Moon. During World War II, the navy had developed a high-chromium, stainless-steel woven fabric that demonstrated high resistance to cuts and abrasions. Even though it was very expensive, it protected Armstrong's back, hands, and feet from punctures. Materials scientists were confident that they could combine these fabrics for use in a spacesuit for the Apollo missions. It would be decades before Museum conservators realized that these materials could not tolerate each other very well over the long term.

TWO HELMETS

Armstrong's A7-L Bubble Helmet, which kept the air inside the spacesuit, was worn at all times when the suit was pressurized, even on the surface of the Moon. *NASM*

It often surprises the public that Neil Armstrong had not one but two helmets. Like the two suits combined into one, Armstrong also wore two helmets when he first stepped on the Moon. Unlike the suit, the helmets were separate, even though they performed analogous functions. The external one visible to the television viewer protected against temperature extremes, solar radiation, and micrometeoroids. Underneath that, he wore a clear, high-strength polycarbonate bubble that attached to his spacesuit via the red neck ring.

NASA engineers James O'Kane and Robert Jones designed the bubble helmet, or Pressure Helmet Assembly. The Air-Lock Corporation manufactured it to provide the astronauts with a clean, unobstructed view of their environment. The bubble helmet allowed the astronauts to see their toes while walking on the surface of the Moon, and it was strong enough to withstand knocks and scratches and hold the 4.3 pounds per square inch of suit oxygen pressure. The blow-molded, clear polycarbonate shell with a molded bayonet base was bonded to the helmet neck ring. Armstrong's neck ring was made from blue anodized aluminum. Later versions of the helmet had red neck rings. Due to design differences, the two versions were not interchangeable.

The only visual obstruction in the bubble helmet is the Vent Pad and Duct Assembly, which is the off-white, mushroom-shaped lining along the back of the head. This was designed to diffuse the oxygen flow around the astronaut's head and diminish the tendency for the helmet to mist up inside when it underwent changes in temperature. On the left side of Armstrong's

Neil Armstrong's spacesuit suit itself is not a single entity but rather a composite of many components, each one a complexly crafted item. The entire complement of equipment, from his underwear out, was called the Extravehicular Mobility Unit (EMU). For simplicity's sake, NASA broke down the EMU into twenty-one units, including two sets of gloves, a liquid cooling garment, external connectors, and all the checklist pockets that an astronaut would need during his EVA. That number did not include the PLSS, which supplied oxygen and cooling water to the astronaut when he was not connected to the spacecraft and transmitted communications and data to the ground.

The part of the EMU that covers the arms, legs, and torso is actually two suits in one that together is called the Integrated Torso Limb Suit Assembly (ITLSA). The ITLSA resulted from NASA's decision to combine the overcoat with the pressure suit for the convenience of the astronauts. It consisted of a predominantly rubber Pressure Garment Assembly (PGA) layer that kept the air inside the suit and restrained the oxygen pressure from ballooning it, as well as a complex Thermal Micrometeoroid

face, a hard, plastic device was attached to the inner neck ring. He could use this to scratch his nose or to compress his nostrils in order to equalize the pressure in his ear tubes. A bit higher and also on the left was a feed port attached to the side of the helmet, with inner and outer halves to assure that it could not be opened by accident. The inner half included a port and gate valve through which he could drink or eat. It was spring-loaded and provided an airtight seal when the probe was removed.

The Lunar Extravehicular Visor Assembly (LEVA) was worn over the pressure helmet, attaching to the necking of the spacesuit, to shield against eye-damaging solar ultraviolet radiation, maintain head and face thermal comfort, and provide an extra shield against micro-meteoroids. Hamilton Standard designed and made the first LEVA designs and prototypes. In parallel, NASA engineers O'Kane and Jones designed a LEVA specifically to fit their bubble helmet. The work was remarkably similar, with O'Kane and Jones conceding Hamilton's point that an addition of thermal insulation was necessary to insulate against temperature changes in the sunlight in their final design. It included an insulated cap with a 24-karat gold-coated visor to filter solar radiation. After testing, heat transfer from the metal prompted them to include a neck-ring cover in 1967. As a result of condensation in the helmet that astronaut Russell Schweickart experienced during Earth-orbital tests on Apollo 9, the LEVA was rede-signed with an outer shell and an attached Integrated Thermal Meteoroid Garment extending down to cover the neck ring. This configuration was the one Armstrong wore on Apollo 11.

Armstrong's Lunar Extravehicular Visor. This is the image of Armstrong's suit that the public most remembers. The EV Visor attached to the bubble helmet as thermal and sun protection. *NASM*

Garment (TMG) that protected Armstrong from the harsh environment of space. Before Armstrong climbed down the ladder of the LM, he had carefully reviewed all the procedures to make certain that his spacesuit was functioning perfectly and that all of the components were in place. His final act to create his personal spacecraft was to disconnect the suit from the LM's oxygen system via the redundant set of blue and red connectors on his torso. The PLSS only began to function in the vacuum of space after the pressure in the spacecraft was reduced to near zero.

Most of what Neil Armstrong wore on the surface of the Moon returned to Earth in the Command Module along with Armstrong and his crewmates. He and Aldrin had tossed their lunar overshoes and backpacks onto the lunar surface a couple hours after the walk. That saved about two-thirds of the weight of the complete suit, increasing the safety margin for the flight of the Lunar Module's ascent stage back to the mothership. While in the safe environment of *Columbia*, the astronauts no longer required the protection of a spacesuit and wore a more comfortable, Teflon fabric in-flight cover garments during much of the remainder of their mission. Once they landed in the Pacific Ocean, a specially trained scuba team handed them isolation suits that were supposed to protect Earth life from any microbes that the Apollo astronauts might have brought back from the Moon. Armstrong, Aldrin, and Collins quickly changed from their spacesuits, which remained inside *Columbia*, which itself was hoisted aboard the aircraft carrier USS *Hornet* and bought to the Manned

CLOCKWISE FROM UPPER LEFT: The Communications Carrier Assembly, nicknamed the Snoopy cap, that Armstrong used during his mission. The formfitting cap kept the communications equipment in place as he turned his head. *NASM*

Armstrong's liquid-cooling garment, which prevented his body from overheating while sealed in the airtight spacesuit. *NASM*

The left intravehicular glove, worn during launch and entry but not on the surface of the Moon. *NASM*

The comfort layer worn underneath the spacesuit to prevent abrasions. *NASM*

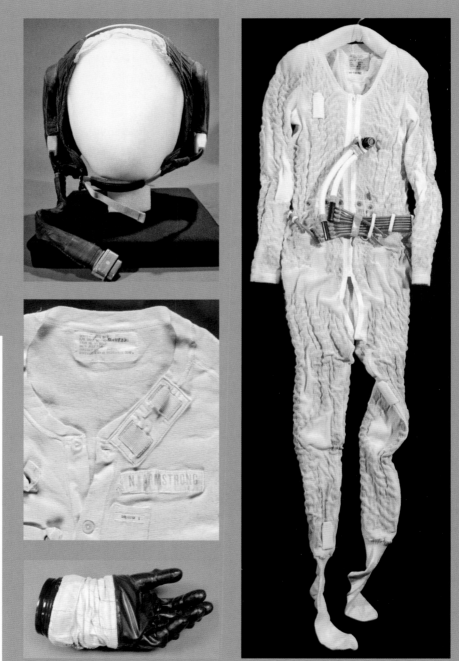

Spacecraft Center in Houston for quarantine, analysis, and treatment.

The technicians who first inspected the Apollo 11 spacesuits were primarily interested in how they had performed and how they might have been damaged during their time in space. The workers in Houston, who were inside a quarantine facility with the astronauts, took close account of the pre- and post-flight leakage values. The vent and pressure time for each was carefully measured. As no seal is perfect, each suit lost a little more than half an ounce of oxygen per minute. Armstrong's suit measured at the lowest end of that pressure loss before flight, yet even afterward it would still have qualified for use in space. The technicians also took samples of lunar dust from the suit, and following the recommendations of Smithsonian conservators, had ILC dry-clean the suit as a preservation measure in its last step as NASA space hardware.

After the Apollo 11 mission ended, NASA sent *Columbia* and its associated items on a tour of the fifty U.S. state capitals, in a van, in 1970. Although all sides had agreed that the artifacts of

LEFT: The Biological Isolation Garment that Neil Armstrong wore to protect the Earth from any Moon germs that he might have brought back from his mission. *NASM*

ABOVE: Armstrong wore this In-Flight Cover Garment Jacket when a spacesuit was not necessary. *NASM*

mankind's first Moon landing would ultimately come to the Smithsonian for display, NASA did not make the formal offer of the Armstrong, Aldrin, and Collins suits to the Museum until April 1971, after the NASA U.S. tour was complete. (Collins was the Apollo 11's Command Module pilot, and coincidentally also the new director of the Museum.) Once Armstrong and Aldrin's suits arrived, they immediately went on display in the National Air and Space Museum's section of the Arts & Industries Building, then moved to the Museum's new building on the Mall in early 1976. Other, smaller components continued their roles as touring artifacts, going on display in Japan, England, and Switzerland as part of the U.S. Information Agency and Smithsonian traveling exhibitions.

Neil Armstrong's spacesuit is currently no longer on exhibit at NASM's *Apollo to the Moon* gallery, although there are plans to return it when a replacement exhibit is built. After thirty years of display with only periodic removals for inspection, conservation concerns prompted staff to put the suit in storage in 2006. It had protected Neil Armstrong on the surface of the Moon, undergone testing after its return, traveled

ILC Industries seamstresses assemble the textile parts of the Apollo spacesuits. *ILC Dover*

It is almost impossible to count the number of hands and minds that formed the hardware that comprises Neil Armstrong's spacesuit. The answer to "Who built it?" is long and complicated, as the suit is made of many individual components. The suit becomes aggregated into a single unit only when worn. The iconic image of a spacesuit in operation shows a machine that was assembled for a mere few hours of its existence.

What people most often perceive as the main component of the Apollo A7-L spacesuit extends from the chin to the feet and from wrist to wrist; that is with the gloves and helmet off. It is actually two spacesuits in one. On the outside is the Thermal Micrometeoroid Garment (TMG), which protects the astronaut from the dangers of extreme temperatures, radiation, and micrometeoroids that travel thousands of miles per hour. On the inside, sustaining life, is the Pressure Garment Assembly (PGA). Though the TMG and PGA were originally conceived as separate suits, NASA and its contractors soon decided that the two should be combined.

At ILC Industries in Dover, Delaware, teams of seamstresses sewed the components of the TMG together on Singer and Brother sewing machines, following rigorous standards for stitch lengths that would have challenged even the finest tailors. Other teams, armed with glue pots, assembled the PGA from pattern pieces of rubber in order to create an airtight, custom-fit air chamber for the astronauts. The seals on the suit also relied on the pressure-sealing zipper that B.F. Goodrich developed and manufactured in 1965 under the guidance of Russell Cooley. Cooley had worked with American pilot Wiley Post to develop the first pressure suit in the 1930s and was the first proponent of the "tomato worm" design that localized air displacement and made the Apollo suits flexible.

The Air-Lock Corporation in Milford, Connecticut, manufactured the characteristic blue and red anodized aluminum connectors that joined gloves, helmet, and hoses in an airtight seal. The company also manufactured the polycarbonate bubble helmet that NASA engineers Jim O'Kane and Bob Jones had designed to give Moon-walking astronauts an unobstructed view down to their feet. Nearby in Windsor Locks, the Hamilton Standard Company manufactured the PLSS backpack that granted astronauts the autonomy to freely explore the lunar terrain. North, in Worcester, Massachusetts, the David Clark Company made the communications carrier headsets that kept Neil Armstrong in contact with Buzz Aldrin, Michael Collins, and Mission Control in Houston.

In less than three years in the late 1960s, all of these companies' staffs, and many more, came together to create this complex machine that would keep Neil Armstrong alive on the Moon. Their work was based on others' efforts to develop aviation pressure suits as well as suits that would preserve life in space. Many hands touched Neil Armstrong's spacesuit before he put it on for the Apollo 11 launch. Many more minds contributed to the technical solutions of designing the first spacesuit that would be worn on the surface of another world.

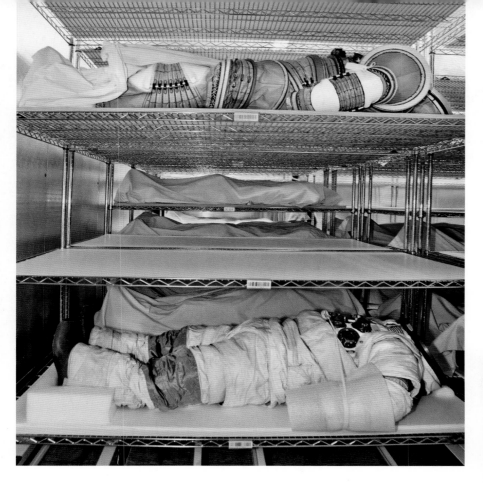

Spacesuit storage at the Steven F. Udvar-Hazy Center. Advanced Apollo suits EX1-A and B1-A are featured in the front. *NASM*

throughout the country, and served as an icon of the Apollo program. But it was beginning to show the detrimental effects of gravity and exposure to sunlight and humidity, as well as the results of the complex interactions between the diverse materials from which it had been assembled. Due the hardening of the rubber pressure bladder and other components, it had become rigid without pressurization.

Conservators, curators, materials scientists, and former spacesuit builders worked together to determine what measures could be taken to preserve this precious artifact short of isolating it into its component materials. Their consensus was that a low, but not cold, temperature and low and stable relative humidity would delay many of the chemical processes that were working against long-term preservation. Since its removal from display, Neil Armstrong's A7-L spacesuit has been in storage at 60°F and 30 percent relative humidity, first at the Museum's Garber Facility and now in the new state-of-the-art storage facilities at the Steven F. Udvar-Hazy Center.

This is not the end of the Neil Armstrong's spacesuit story. To preserve an artifact without the intention of ever again showing it to the public would make it meaningless. Through advanced chemical analysis and imaging, the conservators, material scientists, and curators are exploring new ways to display this suit and continue to preserve it. This way, the public will be able to appreciate this great technological accomplishment for generations to come.

—*Cathleen S. Lewis*

A7-L SPACESUIT
SPECIFICATIONS

MANUFACTURER:
ILC Industries (primary contractor); Hamilton Standard (subcontractor, systems integrator and PLSS manufacturer); B. Welson, Air-Lock, David Clark Company (component manufacturers)
HEIGHT: 66 15/16 in. (167.6 cm)
WIDTH: 20 5/8 in. (50.8 cm)
DEPTH: 11 in. (27.9 cm)
WEIGHT: (empty): 56 lbs. (25.4 kg)

REMAINING ON THE MOON
Lunar Overshoes
LENGTH: 13 in. (33 cm)
WIDTH: 7 in. (17.8 cm)
DEPTH: 7 3/4 in. (19.7 cm)

Personal Life Support System
HEIGHT: 37 in. (94 cm)
WIDTH: 22 in. (55.9 cm)
DEPTH: 10 in. (25.4 cm)
WEIGHT: (empty): 125 lbs. (56.7 kg)

WITH FAR LESS FANFARE than earlier launches that gripped the nation, and without much media coverage, NASA launched Skylab into space on May 14, 1973. True, it was not anything as big as that first small step for man, but it was a significant next step. Assembled largely out of components from the Apollo program, Skylab—America's improvised space station—kept NASA's human spaceflight program in the public eye after the last Apollo Moon landing in December 1972. Its three manned missions set consecutive records for flight duration (twenty-eight, fifty-nine, and eighty-four days), distance traveled, and hours spent on spacewalks. More important, Skylab proved that humans could live, work, and do meaningful research over long periods of time in the weightless environment of near-Earth orbit.

Because the Soviet Union had already put its Salyut into orbit two years earlier, NASA could only tout Skylab as *America*'s first space station. As its name implied, Skylab was really more of an observatory and space-based laboratory than a true space station, the "base camp" to the Moon and Mars that many in NASA sought. Oddly enough, what now seems such an obvious name did not even appear on the list of nearly one hundred suggestions considered by NASA's Project Designation Committee. One might have thought that the navy's earlier SEALAB program would have been the inspiration, but NASA officially credits the name, adopted in 1970, to air force officer Donald Steelman, who was then serving with NASA.

SKYLAB

7

Each three-man Skylab crew included one "scientist-pilot" with a doctorate in medicine, engineering, or science, and all of the astronauts received intensive training in relevant scientific disciplines and observational techniques. Skylab's solar observatory, the Apollo Telescope Mount (ATM), studied solar physics with unprecedented sensitivity and resolution. Its Earth Resources Experiment Package (EREP) pointed a suite of remote-sensing instruments back at the ground. But Skylab's most important scientific contribution may have been in discovering how the human body responds to the conditions of space.

Despite initial skepticism in the scientific community, Skylab's astronauts proved to be resourceful observers, conscientiously running and occasionally repairing complicated experimental apparatus. They enthusiastically responded to unexpected opportunities, such as watching the newly discovered Comet Kohoutek slingshot around the Sun. The vast amount of data collected by Skylab's astronauts on the Sun's magnetic field, on the Earth's oceans and weather patterns, and on human

The third crew bids farewell to Skylab on their final fly-around inspection on February 8, 1974. Note that the main solar panel is missing its twin, while the temporary parasol and twin-pole sunshade are very much still visible. The ATM, with its own solar panels, sits above the orbital workshop. *NASA*

A montage of the three prime Skylab crews made prior to launch. Moving counterclockwise from the top are the astronauts for Skylab 2, 3, and 4. The Apollo Command and Service Modules are at far left, followed by the Multiple Docking Adapter with the ATM on top, the Airlock Module, the Instrument Unit, and the Orbital Workshop. The latter shows its original paint pattern, before the outer shield and the solar wing on this side were ripped off in the launch accident. *NASM*

adaptability to zero gravity, made a strong case that humans had a place in space as scientists as well as pilots and that they could sometimes achieve more than even the most sophisticated satellites. As a template for subsequent research laboratories on the Space Shuttle and the International Space Station, Skylab was invaluable.

After the heady heroics of the Apollo years, Skylab answered, if only temporarily, a pressing question for NASA. How could the experience and resources accumulated in going to the Moon be turned in new directions that would appeal to an American public increasingly disenchanted with the high cost of the space race? Always alert to the threat of future budget contractions, NASA had almost since the beginning of the program been considering creative ways to make use of surplus Apollo hardware and capacity. An orbiting laboratory or observatory emerged quite early as a serious contender, and it had been the subject of several workshops and design feasibility studies. By using off-the-shelf components, an upper stage of a Saturn rocket could be converted into an "orbital workshop" relatively cheaply and quickly, either by retrofitting a spent hydrogen fuel stage after it had reached orbit (the "wet" option) or by outfitting it on the ground and then sending it into orbit (the "dry" option).

The "cluster concept," sketched out by NASA administrator and engineer George Mueller in the summer of 1966, well before the lunar landing, envisioned how different Apollo components—a command and service module, a multiple docking adapter, an airlock module, the ATM, the orbital workshop—could be mixed and matched to fashion a suitable space laboratory, on a bargain-basement budget of $2.5 billion. With Apollo winding down, Skylab had access to one of the last Saturn V rockets, with sufficient power and payload to launch a fully outfitted, hundred-ton Skylab directly into orbit, rather than having to assemble it piecemeal in space. The station itself was a modified third stage with no engines.

Previous NASA missions, being relatively short, had given scant attention to "habitability," meaning accommodations for sleeping, eating, exercise, recreation, and housekeeping. Because Skylab would house crews for weeks or even months at a time, a reasonable degree of both convenience and comfort could very well make the difference between a productive work environment and a stressful one. The astronauts themselves expressed little interest in habitability, as most of them were veterans of the cockpit culture of military pilots. Looking ahead to longer space-flights, NASA administrators sought an outside opinion from noted industrial designer Raymond Loewy.

Reconfigured with Loewy's advice, the cylindrical orbital workshop—forty-eight feet long and twenty-two feet in diameter—had two levels separated by an open metal lattice, a grid of small triangles designed so that the crewmembers could anchor themselves with special cleats attached to their shoes, like a bicyclist to his pedals. The floor and workshop walls had a similar grid, with grips and handholds for moving around in zero gravity. The lower level, essentially the living quarters, had sleeping compartments, a kitchen and dining area or wardroom where the crew could look out of Loewy's "earth observation window," a "waste management compartment" for personal hygiene, a collapsible shower, and an area for biomedical experiments. The larger forward compartment above it had freezers, food and water storage compartments, shielded film vaults, additional room for research equipment, and access to the airlocks. Below the living quarters, an empty, unused liquid oxygen tank served as Skylab's garbage can, complete with a handy, though occasionally tricky, trash disposal airlock. For electrical power, Skylab was to rely on two inter-connected arrays of solar panels, a matching pair of thirty-foot wings set on opposite sides of the orbital workshop, plus four longer, narrower solar panels mounted atop the ATM, like windmill vanes. Each array of silicon solar cells could generate four thousand watts,

Cutaway of the Orbital Workshop showing the main compartments arranged on two decks, including experimental areas, storage facilities, environmental control, and waste disposal systems. The micrometeoroid shield (far right) ripped off during launch. *NASA*

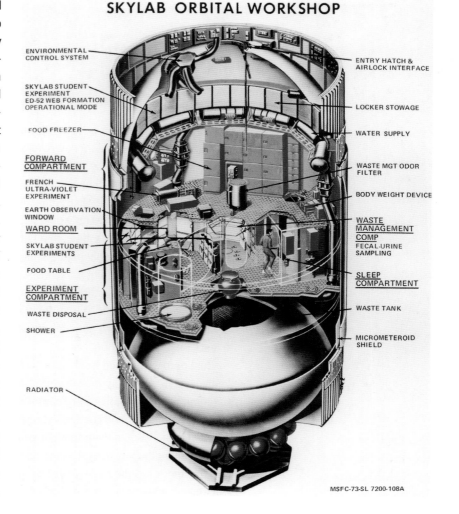

SKYLAB ORBITAL WORKSHOP

ENVIRONMENTAL CONTROL SYSTEM

SKYLAB STUDENT EXPERIMENT ED-52 WEB FORMATION OPERATIONAL MODE

FOOD FREEZER

FORWARD COMPARTMENT

FRENCH ULTRA-VIOLET EXPERIMENT

EARTH OBSERVATION WINDOW

WARD ROOM

SKYLAB STUDENT EXPERIMENTS

FOOD TABLE

EXPERIMENT COMPARTMENT

WASTE DISPOSAL

SHOWER

RADIATOR

ENTRY HATCH & AIRLOCK INTERFACE

LOCKER STOWAGE

WATER SUPPLY

WASTE MGT ODOR FILTER

BODY WEIGHT DEVICE

WASTE MANAGEMENT COMP FECAL-URINE SAMPLING

SLEEP COMPARTMENT

WASTE TANK

MICROMETEROID SHIELD

MSFC-73-SL 7200-108A

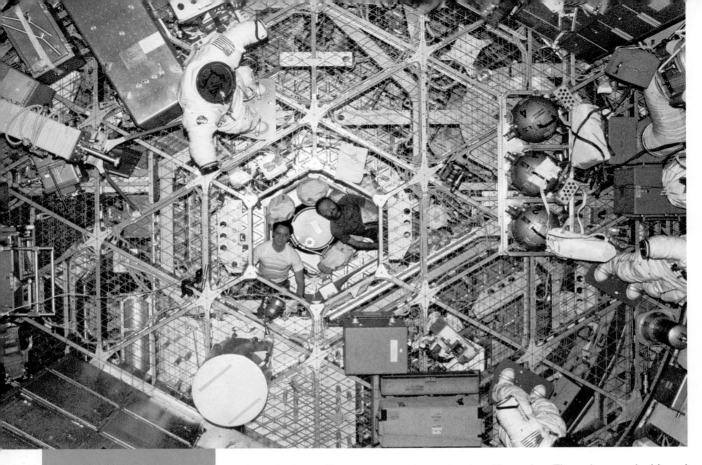

Crewmembers Edward
Gibson and William Pogue
(Skylab 4) look up from the
lower to the upper deck of
the orbital workshop through
the hexagonal opening in the
lattice floor. The spacesuits
on the upper deck stand at the
ready for spacewalks. *NASA*

either for immediate use or to charge banks of batteries. The solar panels, hinged
and carefully folded for launch, would be deployed in orbit.

In space, things easy to do back home on Earth often became frustratingly
difficult, while difficult things became entertainingly easy. Putting on shoes and
socks, for instance, required strenuous exertion of the stomach muscles. Circus-like
acrobatics, on the other hand, could be mastered with ease. The astronauts quickly
discovered also that some ideas that had seemed sensible on the ground never quite
worked in space. The collapsible shower took far too long for convenient set-up
and use, though it did make for some wonderful photo opportunities. The crewmembers each got to choose their individual menu items, an enhancement from earlier
missions, and yet food that was tasty enough back in Houston lost its flavor and
appeal in a low-pressure atmosphere that eliminated the aroma, which is essential
to taste.

Just getting the food from plate to palate in zero gravity took practice. Securing
small items demanded constant vigilance, despite multiple pockets in the astronaut's
suits and the imaginative application of Velcro just about everywhere. Any open
drawer or cabinet could suddenly turn into a "jack-in-the-box"—and then a game of
"hide and seek" when everything floated away. Sooner or later, errant items usually
showed up on the intake screen of the ventilation system. The workshop's low-pressure, oxygen-nitrogen atmosphere limited conversation to fifteen feet or so, making
an intercom essential. On the plus side, weightless sleep had its advantages. Beds

Owen Garriott strapped in and zipped up tight in his sleeping compartment. Each astronaut had a private compartment. In a weightless environment, some chose to sleep upright while others slept upside down, relative to the floor—whatever was most comfortable. *NASA*

Edward Gibson prepares dinner in the wardroom. In zero gravity, chairs would have been pointless. The round Earth observation window, so popular with crewmembers, is closed for the moment. *NASA*

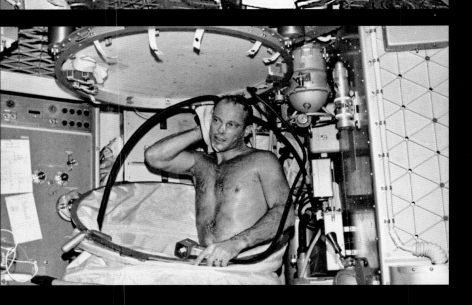

Jack Lousma gives the pop-up shower a try. Even with its specially designed water and air-flow systems and premeasured soap, the shower did not turn out to be a practical alternative to old-fashioned washcloths. *NASA*

DESIGNING A HOUSE IN SPACE

Upon viewing an early mock-up of the Skylab workshop, George Mueller, the head of the NASA manned spaceflight program in the 1960s, said that anyone who lived in the craft for three months would go stir-crazy. As Skylab was meant to set a foundation for long, human forays into the solar system, Mueller insisted that the contractor responsible for the orbital workshop find a habitability consultant to review its interior design.

The contractor, Martin Marietta, hired Raymond Loewy, who is often cited as the father of industrial design and as the creator of the technique of streamlining. Loewy had consulted on a number of transportation-related problems over the course of his career, including the design of President John F. Kennedy's apartment aboard Air Force One. For Skylab, Loewy had to juggle the unique characteristics of confined quarters in zero gravity.

Prior to Skylab, NASA designed its vehicles like aircraft cockpits, where providing basic life support was the guiding principle. Engineers at the NASA Marshall Center in Huntsville, Alabama, saw no need for amenities; instead, they stressed functionality and reliability. Even astronauts scoffed at the suggestion that there should be creature comforts aboard Skylab. They claimed that a living and working space free from petty annoyances would suffice. They had, after all, been trained to carry out experiments under extreme stress and in confined quarters. Loewy believed, however, that a pleasant environment was crucial to effectively live and work in space.

Loewy sought design details from analogous missions, such as long-duration submarine voyages and trips to Antarctica. Finding few, however, he was forced to create a novel blueprint for how a spacecraft should be designed. He mixed practical and aesthetic concerns. For example, while he acknowledged that neutral colors in the workshop might

or cots being pointless, the crew zipped themselves into their sleeping bags, each in his personal compartment, in whatever orientation seemed most comfortable—some head "up," others head "down," and others stretched out hammock-style.

Crews learned to move around Skylab with surprising speed and dexterity, and they found that the fireman's pole and designated handholds and handrails were more hindrance than help. For something more futuristic, they could try the "maneuvering units," jetpacks right out of James Bond, but these were better suited to external use on a full-sized space station rather than in the compact Skylab. The crew had some justifiable complaints about NASA's fashion sense. Despite some ingenious designer touches such as pants that unzipped into shorts, official space wear came in a single shade of golden brown, the only color available in that particular fire-resistant fabric. Because Skylab had no laundry facilities, the crew at least

be psychologically soothing, he ultimately recommended the use of bright colors to provide the astronauts with visual stimulation to prevent boredom and depression. He also situated overhead lights in ways that did not cast eerie shadow patterns on the walls, and he kept noise levels low by muffling air-circulation fans.

Loewy paid attention to the needs of the workshop's distinct areas. He wanted the meal room to be homey, so he suggested that drawings by the astronauts' children be hung on the walls. Loewy also insisted the men have access to privacy and personal space, especially in the waste management area. His most lauded recommendation was the inclusion of a window from which the astronauts could view the Earth from orbit. Engineers at Huntsville complained that the window would weaken the structure of the craft, but with support from Mueller, the vessel was eventually outfitted with a window in the wardroom. Later psychologists confirmed the experience of viewing Earth from space as beneficial for astronauts.

By today's standards, these industrial design recommendations may seem obvious. In the late 1960s, however, Loewy's guidelines for providing a comfortable living and working area in space were both unique and visionary. He realized that if humans were to venture further out into the solar system, they would not simply need to be protected from the hazards of space, but they would also need to feel comfortable in their isolated, confined environment.

Loewy's contributions to the space program were eventually acknowledged by many at NASA, and he further tightened his bond with the agency when he served as the habitability consultant for the Space Shuttle and the International Space Station. He not only laid a foundation for the future design of space stations, he envisioned an approaching reality when men and women could comfortably live and work in the vacuum of space.

Industrial designer Raymond Loewy dons a spacesuit courtesy of NASA. Skylab was sufficiently challenging to draw Loewy out of retirement. *The Hagley Museum and Library*

had the satisfaction of eventually being able to throw it all away. For down time, the men had playing cards (with magnets), Velcro darts (won by trying not to compensate for gravity), books, and recorded music, though window gazing was by far the most popular pastime.

The launch nearly crippled Skylab. The astronauts unexpectedly got an early opportunity to put their training to the ultimate test. As planned, a two-stage Saturn V rocket boosted the unmanned Skylab into its 270-mile circular orbit, where the first crew would rendezvous with and board it. Only a minute into the flight, mission controllers could tell that something had gone wrong, which subsequent telemetry confirmed. Somehow, the shield designed to protect the craft from the impact of micrometeoroids, and simultaneously shelter it from the Sun, had torn off during the launch, shearing off one main solar wing and entangling the other. While

PSYCHOLOGY IN EARTH ORBIT

Astronauts Gerald Carr and William Pogue entertain themselves in the zero gravity environment. Out-of-this-world acrobatics often proved easier than routine daily tasks. *NASA*

The Skylab crews conducted a number of scientific experiments, but the astronauts were also primary objects of research. NASA physicians wanted to investigate whether humans could live in zero gravity for long periods of time without physical impairment. By the end of the 1960s, NASA had plans to construct a large, permanent space station and a shuttle to service it, so the investigation of how astronauts would physically and psychologically respond to long stays aboard Skylab was critical in directing the future of the space program.

Would astronauts succumb to madness in the isolated, confined environment? To assess this possibility, NASA designer Caldwell C. Johnson and NASA experiment developer Robert Bond gathered data about the suitability of Skylab's living and working quarters, which would prove instructive for the design of future spacecraft from a psychological perspective. Crewmembers completed questionnaires and rating forms about the craft's interior design, while environmental measuring instruments and video cameras acquired quantitative data about the conditions aboard Skylab.

Johnson predicted that the Earth-like orientation of Skylab—that is, one in which "up" and "down" have meaning—would be essential for psychological reasons. Astronauts disagreed, however, and asserted that directional markers that have meaning on Earth are superfluous in zero gravity. The astronauts provided insight into other aspects of the vessel's design. The crew cited any device that utilized space in an efficient way as conducive to a successful mission. For example, they appreciated the efficacy of the system used to heat their food trays. In contrast, astronauts complained that their utensils were too small and that the trash accumulation system needed to be reengineered. They also did not see the utility in the provision of a table with upholstered chairs at which to eat their meals, explaining that to take the position of sitting in zero gravity creates stomach muscle strain.

Lapses in psychological comfort aboard Skylab were not instigated by improper habitability standards, but instead by the amount of work the astronauts were asked to perform. Skylab's third crew, which spent eighty-four days in space, complained of excessive workloads and did not handle the stresses of living and working in space as well as had the previous crews.

micrometeoroids did not pose much of an actual threat to Skylab, loss of internal temperature control certainly did. The shield's black and white paint pattern had been carefully designed to balance heat gains and losses. Without it, Skylab's gold foil wrapping, itself a passive heat-control system, would absorb and retain too much solar radiation, sending temperatures inside dangerously high. Moreover, without its main solar panels, Skylab could not generate sufficient electrical power. As a stopgap measure, ground controllers repositioned Skylab to reduce its exposure to

The astronauts were described in contemporary press as being slower and more prone to error than the first and second crews, which heightened any existing tension during the final mission. Members of the Mission Control team eventually devised ways to lessen the stresses and workload put on the astronauts to ensure mission success. Astronauts also found reprieve in scheduled rest days, during which they could enjoy views of the Earth from the wardroom window.

The psychological experiment determined that longer space missions were viable. Design choices generally had a positive impact on the productiveness of the crews or at worst were seen as inessential. It is likely that these Skylab reports will prove invaluable when constructing a vehicle for a mission to Mars, and psychologists since have confirmed the architectural and design decisions pioneered in Skylab. As spaceflight has become an increasingly international endeavor, many "space psychologists" are currently investigating the dynamics of multinational and mixed-gender crews in simulators and in analogous places on Earth, such as stations in the Artic polar desert and underwater habitats. These psychologists recognize that small group studies composed of diverse populations will prove invaluable when planning for long-duration space voyages.

the sun, though at the cost of further reducing the efficiency of the working solar panels. These emergency maneuvers also overheated the attitude control gyroscopes, only adding to a growing list of potentially fatal failures.

While Mission Control worked on diagnosing the problems, the astronauts would have just ten days to prepare to fix them. As a temporary sunshade, NASA engineers devised a "parasol" constructed from spacesuit material—nylon, mylar, and aluminum foil—that could be pushed up through the Sun-facing airlock

and opened like a huge umbrella. Because the material would deteriorate from long-term exposure to ultraviolet rays, the second crew would have to anchor a more permanent "sail" to the ATM and pull it over the parasol. To free the trapped solar panel, NASA engineers modified a standard cable cutter into a space-age pruner. The astronauts practiced repair maneuvers at the Marshall Space Flight Center's neutral buoyancy simulator, a gigantic swimming pool in which they could try out their new tools in something similar to a weightless environment.

The first crew arrived at Skylab on May 25, 1973, assessed the damage, and docked the Apollo spacecraft. With some difficulty—cutting the debris loose from the solar panel took several tries—the crew had Skylab back in business, if slightly worse for wear. The makeshift parasol did its job sufficiently well that within a week the temperature had dropped enough for the crew to move into the workshop. Ground control could then reorient Skylab into its intended attitude, and so regain maximum solar power, though the missing panel could not be replaced. Any questions about the ability of humans to work in space had been answered emphatically. As NASA scientist and noted author Homer Hickam (*October Sky*) reflected: "What I really, *really* like about Skylab is this: When it got into trouble, spacemen armed with wrenches, screwdrivers, and tin-snips were sent up to fix it. No robots, no computers, no remotely controlled manipulating arms, just guys in suits carrying tools."

Would astronauts prove as indispensable for science as they had for do-it-yourself repair? Solar astronomers had long recognized the advantages of putting telescopes in orbit, where the Sun could be studied without the distorting effect of the atmosphere, which absorbs ultraviolet and X-ray solar radiation and often blocks visible light in even the best observing locations. On the ground, the Sun's outer corona is visible only during a rare total eclipse, while space offers an unobstructed view. Orbiting Solar Observatories—the first launched in 1962—had instruments with limited size, resolution, and data storage and transmission capacities. Skylab's ATM, a one-ton behemoth, carried instruments equal to some of the best solar observatories on Earth. The ATM could also take advantage of high-resolution film, because someone would be up there to change the film magazines. Satellites still had to rely on primitive photoelectric sensors and slow radio telemetry. The major challenge for a human solar observatory like Skylab was maintaining extremely precise telescope alignment when even the slightest movement could blur the image. Sophisticated control moment gyroscopes, and observational experience, did the trick.

From the console in the docking adapter, with its two television monitors and joystick, an astronaut could monitor and point the eight ATM telescopes. These had been designed and built by solar astronomers from the Naval Research Laboratory, the Harvard-Smithsonian Center for Astrophysics, the Aerospace Corporation, the High Altitude Observatory, and NASA itself. The console, comparable in complexity to the instrument panel of a large aircraft and outfitted with a more powerful computer than anything on Apollo, could track solar activity in real time, coordinate observations with ground-based astronomers, and quickly zero in on solar flares, coronal holes, and other targets of particular interest.

Working from a minute-by-minute observing schedule prepared by the principal investigators for each instrument, and sent up nightly by mission control, the crewmembers went through their checklists during the solar phase of each orbit, always

Owen Garriott takes a shift at the ATM control console. The crewmembers carried out detailed daily observational instructions sent from the ground. They could also train the telescopic array at particular targets of opportunity, such as solar flares or prominences. *NASA*

The astronauts considered the view from the end of the ATM one of the highlights of the mission. Here, an astronaut swaps out the film magazines, the white box at the end of the boom, containing some of the tens of thousands of images taken by the solar telescopes. *NASA*

alert for promising anomalies that might suddenly alter the plan. Close coordination with a global network of solar observatories and astronomers took full advantage of ATM's capabilities, because it could change targets in a matter of minutes. In addition to manning the console, the crew made a number of adjustments and repairs to the instruments and regularly swapped out the film magazines, some thirty in all, representing 127,000 individual exposures. To do so required a spacewalk to the top of the ATM, a highlight for astronauts on the mission. "To be on the end of the telescope mount, hanging by your feet as you plunge into darkness, when you can't see your hands in the front of your face—you see nothing but flashing thunderstorms and stars—that's one of the minutes I'd like to recapture and remember forever," Jack Lousma, the pilot on the second crew, recalled.

The ATM cluster—two hydrogen-alpha band telescopes (primarily for reference and pointing), two X-ray telescopes, three ultraviolet spectroscopic instruments, and one white light coronagraph—revealed virtually a whole new Sun. Ultraviolet images, some as visually stunning as later Hubble photographs, gave dramatic evidence of the physical properties and processes of the Sun's intensely active chromosphere. X-ray images, in turn, highlighted the structure and behavior of the much hotter corona, identifying coronal holes for the first time as the source of solar wind. One sharp-eyed observer anticipated and then caught on film the full lifecycle of a solar flare. Together, Skylab's solar instruments peeled back the layers of the Sun's atmosphere and its complex interconnected photosphere, chromosphere, transition region, and corona. ATM also recorded and tracked more than one hundred coronal transients, vast bubbles spewed out from the Sun with direct impact on the Earth's ionosphere and magnetic field.

In contrast to ATM, the Earth Resources Experiment Package (EREP) never fully lived up to its advance billing. Added to Skylab at the last minute, as much for political as for scientific purposes, EREP did satisfy calls to give NASA some down-to-earth relevance and responded to the growing environmental movement to which Earth Day in 1970 had given a voice. "I could talk about ATM all day and they'd be polite," one NASA official recounted, "But as soon as I started talking about taking a crop survey, my friends . . . knew what that meant." In other words, the value of environmental surveys were immediately obvious to the average person, whereas solar astronomy was not. EREP, tucked under the opposite side of the docking adapter, mirrored ATM in having a cluster of instruments designed to scan different bands of the spectrum—infrared, microwave, and visible light—each intended to reveal otherwise concealed patterns of cloud cover and weather, geology, surface moisture and vegetation cover, ocean wind and water and conditions, and air and water pollution.

Unlike the close collaboration of ATM, in which teams of principal investigators designed and built their own instruments, NASA designed and built the EREP instruments and only then asked potential users what do with them. NASA solicited more than two hundred proposals from U.S. and foreign scientists, then selected those most likely to yield important scientific results with minimum disruption to the rest of the mission plan. Accommodating EREP did require one fundamental change. So that EREP could survey most of the Earth's populated and arable land, NASA placed Skylab at an orbital inclination of 50° (to the equator) and timed its flight path so that it would make the same pass every five days, for meaningful comparisons of surface features.

Having to juggle the demands of two very different observatories forced some essential compromises in data collection on both sides. EREP pushed up the overall cost by $42 million, an acceptable political price perhaps. EREP's cameras ultimately took forty-six thousand photographs, some striking but few of lasting scientific significance, though they did discover a copper deposit and an oil field. The first Landsat satellite, launched a year before Skylab, returned equally good and more consistent data. Future NASA Earth observatories such as Landsats and Aura, Aqua, and Terra were all unmanned satellites.

Intrigued by the prospect of space-based manufacturing, Skylab's designers also included a small materials processing facility with a furnace, vacuum chamber, and electron beam. The crews grew and doped semiconductor crystals of unprecedented size and purity, and they formed

An enormous solar prominence captured by a Naval Research Laboratory ultraviolet telescope on the ATM. Dwarfing solar flares, prominences are not only visually spectacular—arching a half-million kilometers or more out from the Sun's surface—they provide valuable clues to the Sun's temperature changes and magnetic fields. *NASA*

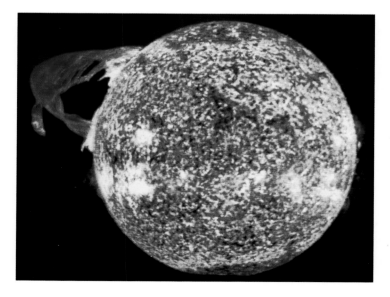

A fisheye view of the Waste Management Compartment. About as user-friendly as it looks, the wall-mounted toilet, with the blue handles, doubled as fecal and urine collector for biomedical experiments. *NASA*

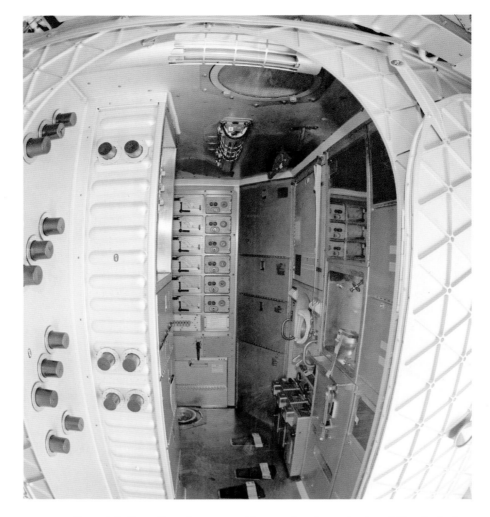

some exotic metal alloys. Despite some evidence of real promise in orbiting factories, the astronomical cost would be an almost certain deal-breaker.

Would humans be able to thrive rather than merely survive on long-duration space flights? Gemini and Apollo astronauts had lost small but nonetheless worrying amounts of body weight, bone calcium, and muscle mass during much shorter missions. How would Skylab's astronauts adjust to prolonged weightlessness? To find out, its crews became both the investigators and the subjects of a battery of sixteen biomedical experiments. Monitoring metabolism, food intake, weight, mineral loss, blood volume and circulation, sleep cycles, balance, and other key physiological factors kept the crews busy. Skylab's high-tech toilet, mounted on the wall of the waste-management unit with vacuum assist and an odor filter—a big upgrade from Apollo—made taking and preserving daily urine and fecal samples relatively easy if never pleasant.

A bicycle ergometer doubled as an exercise machine, with some crewmembers also pedaling with their arms for an upper-body workout. What looked like a rowing machine actually calculated body mass. A cumbersome "lower body negative pressure" instrument, looking perhaps a little too much like an iron lung, simulated gravity to measure how blood pooling in zero gravity (which gave the astronauts a bloated look in the face and torso) affected cardiovascular function.

Space turned out to be a relatively healthy place after all, and the astronauts did not suffer any obvious or irreversible medical conditions. Somewhat surprisingly, the third crew, having spent the longest time in space, recovered the fastest. As in other extreme environments, the human body proved to be remarkably adaptive. For weeks afterward, however, some of the crewmembers' wives noticed that their husbands occasionally had difficulty standing up in the dark, and they displayed the bizarre if understandable habit of expecting everyday objects to hover in midair—with predictable earthly results.

For pure public relations success, nothing could beat the Skylab student projects, culled from 3,700 proposals submitted to the National Science Teacher Association. The twenty-five finalists had the opportunity to showcase their projects for the Skylab team. Nineteen of the experiments actually flew on one of the three missions. The biggest hit was "Web Formation in Zero Gravity," to explore how well spiders could spin a space web. Like the crew, the spiders took some time to get their bearings, but they finally mastered their new environment. So did a colony of bacteria, though not rice seedlings, a possible lesson for future space farmers.

In early February 1974, the third and final crew packed up the remaining film, tapes, samples, and other data and mothballed Skylab. Then, using the thrusters on Command and Service Modules, they boosted Skylab into a higher orbit, hoping to add perhaps a decade to its expected life, then powered it down to await a future mission. For the first time in a decade, no live television crews met the astronauts at splashdown, a reminder of the fickleness of public interest. With Space Shuttle delays making any rescue from it unlikely, and Soviet Salyut successes making it pretty much irrelevant, Skylab had become, with no provision for controlled reentry, a large and potentially lethal piece of space junk. Buffeted by the solar winds that its own instruments had helped explain, Skylab lost altitude more rapidly than predicted, and it broke up over the Indian Ocean and western Australia on July 11, 1979, fortunately without injury or property damage.

Its backup, a flight-ready version, went from NASA storage to the National Air and Space Museum in 1976, where it remains a popular exhibit for visitors trying to imagine life in a zero-gravity environment. Now missing its ATM and EREP (explained by small models), Skylab's significance as a manned orbiting observatory may be somewhat less obvious. The International Space Station—"laboratory, observatory, and factory in space"—has carried Skylab's legacy into the present century, on a scale and with a budget Skylab's designers would have envied, and in a spirit of global collaboration they could scarcely have imagined.

—*Stuart W. Leslie and Layne Karafantis*

The backup Orbital Workshop on display at the National Air and Space Museum. Its strictly vertical orientation gives a somewhat misleading sense of habitability in space. The gleaming gold finish did not long survive the rigors of space. *NASM*

SKYLAB
SPECIFICATIONS

MANUFACTURER:
  McDonnell Douglas
LENGTH: 48 ft. (14.6m)
DIAMETER: 22 ft. (6.7m)
WIDTH ACROSS SOLAR ARRAYS:
  90 ft. (27m) tip-to-tip wingspan
WEIGHT: 78,000 lbs. (35,380 kg)

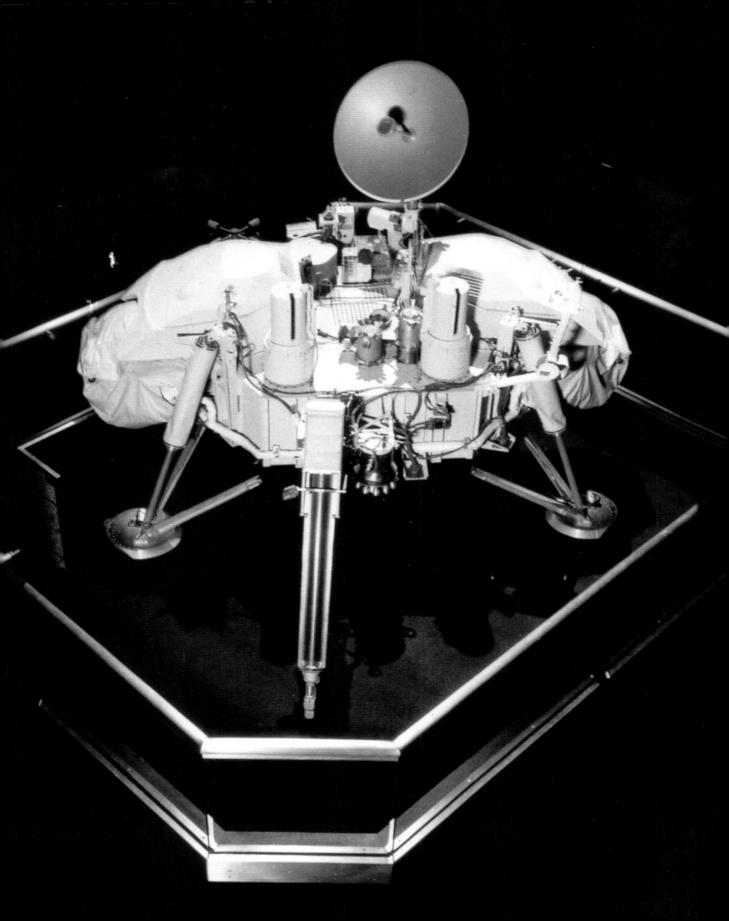

ON JULY 20, 1976, the Viking 1 lander became the first robotic spacecraft to reach the surface of Mars and transmit scientific data. The engineers and scientists who sent it to Mars—as well as its twin lander, Viking 2, which reached the surface two months later—had faced more than a decade of difficult politics, budgetary battles, and technological challenges, achieving this success in the midst of the bicentennial celebration of 1776. Moreover, these soft landings by Viking 1 and Viking 2 represented the culmination of a series of U.S. missions to explore the planet Mars that had begun in 1965 with a flyby by Mariner 4 and continued with several missions thereafter.

These landers, as well as two orbiters that were also a part of the Viking program, utterly transformed scientific knowledge about the red planet. Mars has long held a special fascination for humans—above all because of the possibility that life might exist there. No one in America was more identified with Mars than astronomer Percival Lowell, who built the Lowell Observatory near Flagstaff, Arizona, in the later nineteenth century to study the planet. He argued that Mars had once been a watery planet and that the topographical features he called canals had been built by intelligent beings.

In 1908, Lowell published *Mars as an Abode of Life*, which was based on his extensive visual observations and an exceptionally active imagination. He offered a compelling portrait of a dying planet, whose inhabitants had constructed the vast

THE VIKING LANDER /8

OPPOSITE: The Viking lander consisted of a six-sided aluminum base supported by three legs. NASA attached the instrumentation to the top of the base, elevated above the surface by the extended legs. Power was provided by two radioisotope thermal generator (RTG) units containing plutonium 238 on opposite sides of the lander base and covered by wind screens that provided 30 watts of continuous power at 4.4 volts. Instruments studied the biology, chemical composition (organic and inorganic), seismology, magnetic properties, appearance, and physical properties of the Martian surface and atmosphere. The lander had an arm to collect soil samples, which also had a temperature sensor and magnet on its end. The gas chromatograph mass spectrometer was a state-of-the-art experiment to characterize the chemical elements of Mars. A meteorology boom measured wind direction and velocity, atmospheric pressures, and temperature. Two 360-degree cylindrical scan cameras recorded the visible light features of the surface, and both high- and low-gain communication systems relayed data to Earth and received commands from scientists. *NASM*

This map from Lowell's *Mars as an Abode of Life* (1908) shows the canals that he thought existed on the Martian surface. *NASM*

Homer E. Newell, NASA associate administrator of space science throughout the 1960s. He was responsible for starting and sustaining the Mars exploration program. *NASA*

irrigation system of canals to distribute water from its polar regions to the population centers nearer the equator. Despite its popular appeal, many astronomers refused to accept Lowell's theory. They could not confirm his observations, and many soon concluded that he had been fooled into seeing the canals.

Lowell's observations gave rise to the Martian myth, one of the most powerful ideas motivating human curiosity about the solar system during the twentieth century. Many people genuinely expected explorers to find life on Mars. Magazine illustrations commonly portrayed the planet with a network of canals. Editors at *Life* magazine informed readers in 1944 that the canals served to irrigate patches of vegetation "that change from green to brown in seasonal cycles." Willy Ley, one of the most popular science writers of that time, assured readers of a 1952 issue of *Collier's* magazine that primitive plant life "like lichens and algae" surely existed on Mars. Where plants exist, Ley added, animals must have followed. Walt Disney, in a widely viewed 1957 television broadcast, showed animated drawings of flying saucers skimming over fields of Martian plants and animal life.

As late as the early 1960s, some scientists still argued that vegetation changed colors through planetary seasons on Mars. Gerard P. Kuiper, using a powerful ground-based telescope at the McDonald Observatory, claimed he saw a "touch of moss green" on Mars, sending fellow astronomers searching for a chlorophyll signature using spectroscopic analysis. Some claimed to find it, and for several years scientists reported that vegetation on the planet, probably lichens or some other type of plant life, changed with the seasons. These vivid colors finally proved illusory, and the greens and blues reported turned out to be visual tricks when neutral-toned areas were surrounded by yellow-orange dust storms.

The idea of intelligent life on Mars did not fully depart American popular conceptions until Mariner 4 flew within 6,118 miles of Mars on July 15, 1965, taking twenty-one close-up pictures. They dashed the hopes of many that life might be present on the red planet. These first closeup images of Mars showed a cratered, lunarlike surface. They depicted a Mars without artificial structures and canals, nothing that even remotely resembled a pattern that intelligent life might produce. *U.S. News and World Report* announced that "Mars is dead." Even President Lyndon Johnson pronounced that "life as we know it with its humanity is more unique than many have thought."

Notwithstanding this development, NASA scientists insisted on pursuing an aggressive Mars exploration program. Based on recommendations from planetary scientists, NASA's Office of Space Science formulated a $2 billion program (in 1960s dollars) to search for life on Mars. Known at that time as Voyager, it is not to be confused with Voyagers 1 and 2, which later went to the outer planets. To fund this effort, Homer Newell, leading the NASA science program, canceled plans for missions to other planets. Newell believed that a preponderance of opinion from the scientists supported the Mars program. He was wrong. While the majority of scientists probably supported the mission, a vocal minority thought it too risky and expensive. A public dispute spilled into the American Capitol.

In the summer of 1967, because of conflicting testimony before Congress, a shortage of funds caused by the Vietnam War and Great Society programs, and infighting among space scientists, NASA was forced to cancel the Voyager Mars program. That fall, frustrated by this internal strife, NASA Administrator James E. Webb stopped all work on new planetary missions until scientists could agree on a way forward. Thereafter, they went to work and hammered out an acceptable planetary program for the 1970s.

The space science community learned a hard lesson in practical politics from the Voyager fiasco, as well as infighting over the proposed planetary Grand Tour in 1971. Most important, they learned to resolve their differences in internal discussions, not in public complaints to the media or in testimony before Congress. They also learned that while strong scientific support could not necessarily guarantee political support for a mission, lack of agreement among space scientists would certainly ensure a program's demise.

Retrenched and restructured, a succession of stunning missions followed

A "real-time data translator" machine converted Mariner 4 digital image data into numbers printed on strips of paper. Too anxious to wait for the official processed image, NASA engineers attached these strips side by side to a display panel and hand colored the numbers like a paint-by-numbers picture. It showed a cratered surface that suggested that Mars was less habitable than previously thought. *NASA/JPL*

CARL SAGAN

Carl Sagan played a leading role in Mars exploration from its earliest years, and no one was more important as a space advocate between the 1970s and his death in 1996. His powerful intellect, striking charm, and enormous charisma enabled him to become the public's intellectual in matters of space science and exploration. But he also engaged in strictly scientific pursuits. He worked with NASA beginning in the 1950s, taught Apollo astronauts about lunar science, and labored as a scientist on the Mariner, Viking, Voyager, and Galileo expeditions to the planets. He helped solve the mysteries of the high temperature of Venus (a massive greenhouse effect), the seasonal changes on Mars (windblown dust), and the reddish haze of Titan (complex organic molecules). Fittingly, Asteroid 2709 Sagan was named for him.

For the Viking program, Sagan served on the landing site selection committee, always a lengthy, involved, hotly contested process in which scientists and engineers with broadly divergent views wrestle over the best place to set the lander down on the Martian surface. Depending on their individual specialties, scientists would argue for landing near certain features. Depending on other mission parameters and perceived riskiness of terrain, engineers would accept or resist those proposals. Over time, this committee arrived at a ranking of landing sites. Those engaged in this deliberation served as unsung heroes in the process of conducting the Viking program.

Sagan made a critical contribution to the landing site selection committee by championing radar astronomy as a means of determining more about soil composition, surface features, and rocks on proposed landing sites. But because this approach required a selection of a limited number of sites for intensive radar investigation, other scientists were hesitant to limit their options in this way. Sagan stepped in during February 1973 to press the issue, in part because of a fear that there might be deep fissures of dust on Mars as well as the possibility that "quicksand" had caused the failure of a Soviet lander, Mars 3, in 1971. He emphasized that at "a recent landing site working group meeting we were all entertained to see a Viking lander sinking up to its eyebrows. . . . While a similar suggestion that lunar landing spacecraft would sink into surface dust has proved erroneous, it by no means follows that quicksand is not a hazard for Mars."

Sagan insisted that additional research be undertaken to understand radar data already returned, that analogue studies be undertaken on Earth to help understand what might be seen during Mars radar imaging, and that such facilities as the great radio telescopes at Arecibo, Haystack, and Goldstone be made available to collect extensive Mars imagery during the 1973 and 1975–76 period when the planet was closest to Earth. He added that American scientists "have a very impressive Mars mapping capability, which should be exploited to the fullest." Accepting this argument, NASA established a Viking radar study team and redoubled efforts

Carl Sagan with the Viking lander mock-up in Death Valley, California, on October 26, 1980. *NASA/JPL*

to learn more about the regions of the Martian surface that were potential landing sites for Vikings 1 and 2. It proved an important transition in the site selection process and helped to ensure the success of these spacecraft.

Sagan also served on the science team during the missions themselves, and he was excited by the prospect of finding life on Mars. For years, he had speculated about the possibilities of life beyond Earth, and of all the places in the solar system Mars seemed to hold great promise for finding it. Because the landers found no evidence of life, Sagan and others were disappointed. However, he refused to be deterred in his optimism that it might ultimately be found on the red planet, possibly in Martian microenvironments where life might still survive in small oasis-like areas. Until the end of his life, Sagan still held out hope that robotic probes to the planets might uncover some evidence for life beyond Earth.

As a Pulitzer Prize–winning author, Sagan wrote many bestsellers, including *Cosmos*, which became the best-selling science book ever published in English. This work accompanied his Emmy- and Peabody award–winning television series, seen by five hundred million people in sixty countries. At the time of his death on December 20, 1996, he served as the David Duncan Professor of Astronomy and Space Sciences and director of the Laboratory for Planetary Studies at Cornell University. Sagan's last book, *The Demon-Haunted World: Science as a Candle in the Dark*, was released by Random House in March 1996. Appropriately, this book offered science-based analysis of purportedly supernatural events. He also cowrote the acclaimed film *Contact*, based on his powerful novel of humanity's encounter with extraterrestrial life, the search for which he dedicated his life.

Image from the Viking 1 lander looking northeast toward the rock nicknamed "Big Joe." The largest rock visible near the lander, it is about two meters wide. *NASA/JPL*

throughout the 1970s, even as budgetary pressures and reduced political support remained. These included not only the Viking program for Mars exploration but also Jupiter and Saturn missions by Pioneers 10 and 11 and a new Grand Tour of all of the gas giants in the solar system by Voyagers 1 and 2.

In this context, NASA developed spacecraft in the latter 1960s that laid the groundwork for an eventual landing on the red planet. The flybys of Mariners 6 and 7 verified the Moonlike appearance of Mars, but scientists also determined from these missions that volcanoes had once been active on the planet, that the frost observed seasonally on the poles was made of carbon dioxide, and that huge plates indicated considerable tectonic activity in the planet's past. The Mariner 9 Mars orbiter proved even more interesting in November 1971 when it found the remains of giant extinct volcanoes dwarfing anything on Earth. The largest, Mons Olympus, measured three hundred miles across at the base, with a crater in the top forty-five miles wide. Rising twenty miles from the surrounding plane, Mons Olympus was three times the height of Mt. Everest. Pictures also showed a canyon, Valles Marineris, 2,500 miles long and 3.5 miles deep. Later, Mariner 9 imaged meandering "rivers" which suggested that some ancient past water had probably flowed on Mars.

Suddenly, Mars fascinated scientists, reporters, and the public again, in part because of the possibility of life that might once have existed there. With the success of the Mariner flybys, NASA was able to gain approval for a proposed $2 billion project named Viking to soft-land on the red planet. The Viking mission that emerged in the early 1970s consisted of two identical spacecraft, each with a lander and an orbiter. Launched in 1975 from the Kennedy Space Center in Florida, Viking 1 spent nearly a year cruising to Mars, placed an orbiter in operation around the planet, and landed on July 20, 1976, on the Chryse Planitia (Golden Plains), with Viking 2 following in September 1976. These were the first landings on another planet in the solar system that lasted long enough to deliver useful scientific data.

NASA's Langley Research Center in Hampton, Virginia, managed the Viking project from its inception in 1968 until April 1, 1978, when the Jet Propulsion Laboratory assumed the task of operating the spacecraft. Martin Marietta Aerospace in Denver, Colorado, developed the landers. NASA's Lewis Research Center in Cleveland, Ohio, had responsibility for the launch vehicles. JPL's initial assignment was the development of the orbiters, tracking, and data acquisition systems. Both spacecraft rode into space atop Titan III-E rockets using Centaur third-stages.

The primary objective of the Viking program was to undertake a sustained investigation of the geophysical properties of Mars. From orbit, the spacecraft performed a reconnaissance of the surface, locating possible landing sites and other physical features of interest on the planet. The two landers, communicating to Earth via the orbiters, continuously monitored physical conditions on the ground.

With a single exception—the seismic instruments—the scientific return from the expedition was spectacular. Unfortunately, the seismometer on Viking 1 did not work after landing, and the seismometer on Viking 2 detected only one event that may have been seismic. On the other hand, the two landers continuously monitored

This first panoramic view by Viking 1 from the surface of Mars in 1976 depicts an out-of-focus spacecraft component toward left center, the housing for the Viking sample arm that has not yet been deployed. Parallel lines in the sky are an artifact and are not real features. However, the change of brightness from horizon toward zenith and toward the right (west) is accurately reflected in this picture, taken in late Martian afternoon. At the horizon to the left is a plateau-like prominence much brighter than the foreground material between the rocks. The horizon features are approximately three kilometers (1.8 miles) away. At left is a collection of fine-grained material reminiscent of sand dunes. The dark sinuous markings in left foreground are of unknown origin. Some unidentified shapes can be perceived on the hilly eminence at the horizon toward the right. A horizontal cloud stratum can be made out halfway from the horizon to the top of the picture. At left is seen the low-gain antenna for receipt of commands from the Earth. The projections on or near the horizon may represent the rims of distant impact craters. In right foreground are color charts for lander camera calibration, a mirror for the Viking magnetic properties experiment, and part of a grid on the top of the lander body. At upper right is the high-gain dish antenna for direct communication between landed spacecraft and Earth. *NASA/JPL*

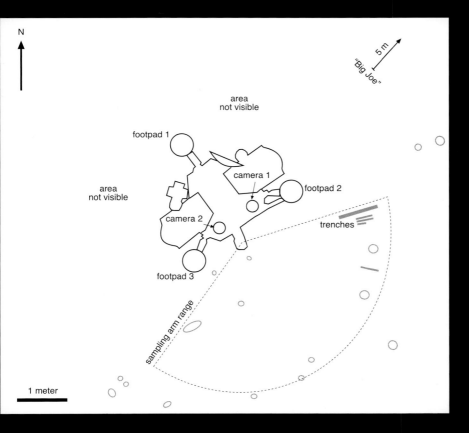

The Viking 1 lander was the first successful arrival on the surface of Mars. Viking 1 transmitted images and meteorological data to Earth for longer than six years. The Viking landers also carried seismometers and a sampling arm. The arm could reach to the surface to scoop soil and return it to a miniaturized laboratory on the lander. This map shows the Viking 1 landing site. Prominent rocks are shown on the surface, along with trenches dug by the sampling arm. In 1981, ownership of the Viking 1 lander was transferred to the Smithsonian National Air and Space Museum. Renamed the Thomas A. Mutch Memorial Station, it is the most distant piece in the museum's collection. It was named after Thomas Mutch, the former lead scientist of the Viking lander imaging experiment. *NASM/NASA*

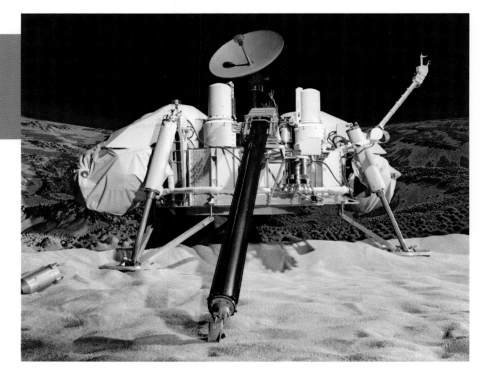

This "glamour" photograph of the Viking lander from 1976 displays its major features, including its scoop and chemistry instruments. *NASA/JPL*

Viking 2 landed on September 3, 1976. Identical to the Viking 1, it operated on the surface for three years and seven months. Both Viking landers utilized a radioisotope power supply. Powered by the radioactive decay of plutonium, these power supplies allowed the landers to operate for many years without relying on solar panels. *NASM/NASA*

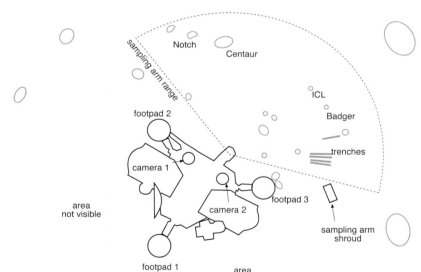

weather at the landing sites and found exciting cyclical variations as well as an exceptionally harsh climate. Atmospheric temperatures at the more southerly Viking 1 landing site, for instance, were only as high as 7°F at midday, but the predawn summer temperature was −107°F. The lowest predawn temperature was −184°F, about the frost point of carbon dioxide. The project also observed the Martian winds, finding that they generally blew more slowly than expected.

The Viking project's primary mission ended on November 15, 1976, eleven days before Mars's superior conjunction (its passage behind the Sun), although the Viking 1 lander continued to operate for six years after first reaching Mars. Its last transmission reached Earth on November 11, 1982. The Viking 2 lander had earlier been turned off on April 11, 1980, after its batteries had failed, and the Viking 1 and 2 orbiters were shut down in August 1980 and July 1978, respectively, after their attitude control propellants ran low. NASA finally closed down the overall mission on May 21, 1983. NASA Chief Scientist James Garvin summed up this mission: "Chryse Planitia is an interesting place. . . . Long ago—perhaps billions of years—it was the dumping ground for five wide outflow channels apparently carved by flowing water." He went on, "Viking 2 landed at 48° north latitude near the Mie crater. . . . It was a very different environment from the flood plains of Chryse." Moreover, Viking 2 had one leg on a rock that tilted the spacecraft at an 8° angle. Regardless, it functioned effectively until April 11, 1980, when its batteries failed and the mission control ceased communication with it.

Although the three biology experiments on the landers discovered unexpected and enigmatic chemical activity in the Martian soil, they provided no clear evidence for the presence of living microorganisms in the soil near the landing sites. A false positive excited scientists for a short time, when the spacecraft mistook peroxides or other oxidants as biological material because of calibration issues. "The landers revealed an alien world with sterile soil and eerie salmon-pink skies," a NASA statement said. "No plants swayed in the breeze. No animals scurried from rock to rock." According to scientists, Mars was self-sterilizing. "On the bright side," opined one wit, "there were no hostile aliens either." Scientists learned that the combination of solar ultraviolet radiation that saturated the surface, the extreme dryness of the soil, and the oxidizing nature of the soil chemistry had prevented the formation of living organisms in the Martian soil.

The failure to find any evidence of life on Mars, past or present, devastated the optimism of scientists involved in the search for extraterrestrial life. These missions led first to the abandonment by many scientists of the hope that life might exist elsewhere in the solar system. Planetary scientist and JPL director Bruce Murray complained at the time of Viking about the lander being ballyhooed as a definite means of ascertaining whether or not life existed on Mars. The public expected to find it, and so did many of the other scientists involved in the project. Murray argued that "the extraordinarily hostile environment revealed by the Mariner flybys made life there so unlikely that public expectations should not be raised." Murray believed that the legacy of failure to detect life, despite the billions of dollars spent and a succession of overoptimistic statements, would spark public disappointment. Murray

MARS LANDINGS

Since the beginning of the space age, there have been fifteen attempted soft landings on the surface of Mars and three on its satellite Phobos, many of which never made it to the target or crashed. The Soviet Union carried out two attempted landings in 1971, Mars 2 and 3, but the first lander crashed and the second returned only twenty seconds of data before failing on the Martian surface. Regardless, these became the first human-built artifacts to reach the surface of Mars. Another landing attempt took place in 1973 when the Soviet Union dispatched both Mars 6 and Mars 7 to the red planet. Mars 6 transmitted signals during the descent but crashed while attempting to land. Mars 7 failed to rendezvous with Mars and went into a solar orbit without accomplishing its mission. A major contribution came in 1976 with the successful landings of Vikings 1 and 2 by the United States.

After the Vikings, no other landers were successful in reaching the Martian surface or Phobos (first attempted by the Soviet Union in 1988) until 1997, when Mars Pathfinder opened the modern age of Martian exploration by landing on the planet. Thereafter, four additional landers have successfully made it to the surface of the red planet and have reshaped humanity's understanding it.

It might be expected that landing failures would have been common early in the space age, and that greater success should come with time, experience, and more sophisticated technology. This is generally the case, but unfortunately in the last fifteen years there have been two U.S. Mars landing mission failures . . . Mars Polar Lander and Deep Space 2 in 1999. The Russian Space Agency has had very little success. Mars 96, an orbiter carrying two landers, and Phobos-Grunt, an attempt to return a sample from Phobos, were both lost due to booster failures, in 1996 and 2011 respectively. Successfully reaching the surface of the red planet or its

was at least partly right. NASA did not return to Mars for two decades. The Viking Program's chief scientist, Gerald A. Soffen, commented in 1992: "If somebody back then had given me 100 to 1 odds that we wouldn't go back to Mars for 17 years, I would've said, 'You're crazy.'"

A second reaction, never accepted by scientists, found a powerful public life. Some asserted that a corrupt federal government, and its mandarins of science, had found evidence of life beyond Earth but was keeping it from the public for reasons ranging from stupidity to diabolical plots. NASA had to respond to these charges repeatedly thereafter. This issue first arose on July 25, 1976, when the Viking 1 orbiter took an image of the Cydonia region of Mars that looked like a human face. All evidence suggests that this was the result of shadows on the hills, and Gerry Soffen said so at a press conference, but some refused to accept this position. The "face" has remained

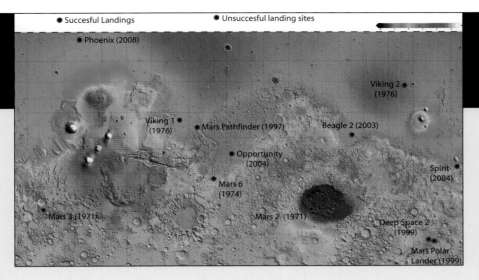

Succesful Landings   Unsuccesful landing sites

Phoenix (2008)

Viking 2 (1976)

Viking 1 (1976)

Mars Pathfinder (1997)

Beagle 2 (2003)

Opportunity (2004)

Spirit (2004)

Mars 6 (1974)

Mars 3 (1971)

Mars 2 (1971)

Deep Space 2 (1999)

Mars Polar Lander (1999)

This map shows the location of American and Soviet landing attempts on the surface of Mars. None of the Soviet Mars landers operated long enough to return useful data despite reaching the surface. Also shown is the Mars Polar Lander which crashed in the south polar region. Phoenix, a similar mission in 2008, successfully landed in the north polar region. *NASM/NASA*

moons has proven a task not without difficulties, yet the prize of scientific knowledge continues to spur significant efforts. There is no dearth of plans for continued exploration using landers, rovers, and flying machines that might operate in the thin Martian atmosphere.

Mars and Phobos Landing Scorecard

Nation	Successful Missions	Partially Successful Missions	Unsuccessful Missions
USA	6	0	2
USSR/Russia	0	1	8
European Space Agency	0	0	1
Total	6	1	11

a sore point, with Soffen being asked about it many times over the years. Always, he stated it was not the remnant of some ancient civilization but was a natural feature lit oddly in this one image but not in any others. As NASA stated officially in 2001, "The 'Face on Mars' has since become a pop icon. It has starred in a Hollywood film, appeared in books, magazines, radio talk shows—and even haunted grocery store checkout lines for 25 years. Some people think the Face is *bona fide* evidence of life on Mars—evidence that NASA would rather hide, say conspiracy theorists. Meanwhile, defenders of the NASA budget wish there *was* an ancient civilization on Mars."

The Viking landing mission had been predicated on the belief that microbial life would be found in the Martian surface. Failure to discover any microbes proved devastating to the cause of life on Mars, but some still clung to the hope that they had looked in the wrong locations or had designed the experiments in a way that

This image was taken by the Sojourner rover's right front camera on the thirty-third day of the mission in 1997. The rock in the foreground, nicknamed "Ender," is pitted and marked by a subtle horizontal texture. The bright material on the top of the rock is probably wind-deposited dust. The Pathfinder Lander is seen in the distance at right. The lander camera is the cylindrical object on top of the deployed mast. This area is considered part of an ancient flood plain. *NASA/JPL*

Viking found no evidence of surface life, or even life that might live at the depths that the lander could dig on the Martian surface. As it turns out, that should not have been surprising because surface dwellers are probably rare. On Earth, most of the biomass lives below the planetary surface in the soil or the oceans. Regardless of negative results from the Viking landers, this fact offered something for scientists to cling to as they considered future exploration of the red planet.

The failure to find evidence of life on Mars devastated the optimism present for astrobiology in an era of great expectations at the time of the Viking landings on Mars. While some scientists were discouraged at first, they grew more optimistic because of two major events in 1996–97. First, in August 1996 a team of NASA and Stanford University scientists announced that a Mars meteorite found in Antarctica contained possible evidence of ancient Martian life. When the 4.2-pound, potato-sized rock (identified as ALH84001) was formed as an igneous rock about 4.5 billion years ago, the scientists believed that Mars was much warmer and probably contained oceans hospitable to life. Then, about fifteen million years ago, a large asteroid hit the red planet and launched the rock into space, where it remained until it crashed into Antarctica around 11,000 BCE. Scientists presented three suggestive, but far from conclusive, pieces of evidence asserting that fossil-like remains of Martian microorganisms, which dated back 3.6 billion years, might be present in ALH84001. While there is no consensus on the truth of these findings, and ultimately most scientists have rejected them, they did lead to added support for an aggressive set of missions to Mars to help discover the truth.

Second, data from Mars Pathfinder in the summer of 1997 demonstrated that the lander rested on an ancient flood plain on Mars and that water had once flowed freely on the surface. Thereafter, the strategy for much of Mars exploration has been built upon the motto "Follow the Water." In essence, this approach noted that life on Earth is built upon liquid water and that any life elsewhere would probably have chemistries built upon these same elements.

did not yield useful results. But the idea did not completely die. Moreover, Viking's substantial cost deterred efforts to launch another Mars lander for twenty years. During that period, the technology of robotic flight advanced considerably. By the 1990s, scientists had begun to approach the issue of life on Mars in another way, modifying earlier conceptions of a solar system containing life beyond Earth. They admitted that liquid water on the surface of Mars would either freeze or evaporate almost immediately, and that the atmosphere was also almost waterless. Even so, they asserted that features seen from space looked like they had been carved by rivers and fast-flowing floods.

Accordingly, to search for life on Mars, past or present, NASA's strategy must be to follow the water. If scientists could find any liquid water on Mars, probably only deep beneath the surface, the potential for life to exist would also be present.

These ideas were confirmed by space probes at the planet since the latter 1990s. The spacecraft that opened this possibility was Mars Global Surveyor, which reached the planet in 1998 and inaugurated a new era of scientific study. In an exciting press conference in June 2000, astronomer Michael Malin discussed his analysis of imagery from the spacecraft, indicating that more than 150 geographic features all over Mars were probably created by fast-flowing water. He suggested that there might actually be water in the substrata of Mars, and our experience on Earth has indicated that where water exists, life as we understand it exists as well. Operating for several years, Mars Global Surveyor continued to send back views of the Martian surface that showed evidence of dry riverbeds, flood plains, gullies on Martian cliffs and crater walls, and sedimentary deposits that suggested the presence of water flowing on the surface at some point in the history of Mars.

At present, most planetary scientists believe that it is unlikely that complex life forms could have evolved on Mars because of its extremely hostile environment. The stories of "advanced civilizations," as proposed by Percival Lowell, or "little green men," are just that—stories. But many scientists believe there is sufficient evidence to think that microscopic organisms might once have evolved on the planet when it was much warmer and wetter billions of years ago. There are even a few scientists who would go further and theorize that perhaps some water is still present deep inside the planet. In that case, simple life forms might still be living beneath Mars's polar caps or in subterranean hot springs warmed by vents from the Martian core. These simple life forms might be Martian equivalents of single-celled microbes that dwell in Earth's bedrock. Planetary scientists and astrobiologists are quick to add, however, that these are unproven theories for which evidence has not yet been discovered.

The last decade of the twentieth century brought new possibilities, as data from the ALH84001 meteorite and the mission of the Mars Pathfinder spacecraft showed the potential of past liquid water flowing freely on the planet. With water as the fundamental building block of life, the search for life on Mars entered a new arena: a scaling back to past life now extinct on a dead world. Scientists may yet find it. Certainly, the evidence of past water on the planet is compelling. If fossils of prehistoric Martian creatures are found, the discovery would hold important—even profound—implications for humanity. But with every new piece of evidence about the lack of life on the planet, many scientists still do not abandon hope; instead, they modify the desire just enough to continue their search for life beyond Earth.

The most exciting discovery on Mars is the now well-accepted consensus that it was once a watery planet that held the building blocks of life. Mars remains an inviting target, all the more so because of extraordinary findings from Mars Global Surveyor, which orbited and mapped the Martian surface from March 1998 to January 2007. It imaged gullies on Martian cliffs and crater walls, suggesting that liquid water has seeped onto the surface in the geologically recent past. This was confirmed by Mars Odyssey 2001, another NASA orbiter, which found that hydrogen-rich regions are located in areas known to be very cold and where ice should be stable.

This relationship between high hydrogen content with regions of predicted ice stability led scientists to conclude that the hydrogen is, in fact, in the form of ice. The ice-rich layer is about two feet beneath the surface at 60° south latitude, and it gets to within about one foot of the surface at 75° south latitude. This evidence suggests that the planet was once significantly more habitable than it is today. Of course, it remains

RIGHT: The boulder-strewn field of red rocks reaches to the horizon nearly two miles from Viking 2 on Mars's Utopian Plain. Scientists believe the colors of the Martian surface and sky in this photo represent their true colors. Fine particles of red dust have settled on spacecraft surfaces. The salmon color of the sky is caused by dust particles suspended in the atmosphere. Color calibration charts for the cameras are mounted at three locations on the spacecraft. Note the blue star field and red stripes of the flag. The circular structure at top is the high-gain antenna, pointed toward Earth. Viking 2 landed on September 3, 1976, some 4,600 miles from its twin, Viking 1, which touched down on July 20. *NASA/JPL*

LEFT: The Viking 1 Orbiter spacecraft photographed this region in the northern latitudes of Mars on July 25, 1976, while searching for a landing site for the Viking 2 lander. The speckled appearance of the image is due to missing data, called bit errors, caused by problems in transmission of the photographic data from Mars to Earth. Bit errors comprise part of one of the "eyes" and "nostrils" on the eroded rock that resembles a human face near the center of the image. Shadows in the rock formation give the illusion of a nose and mouth. Planetary geologists attribute the origin of the formation to purely natural processes. The feature is 1.5 kilometers (one mile) across, with the Sun angle at approximately 20°. The picture was taken from a range of 1,873 kilometers (1,162 miles). *NASA/JPL*

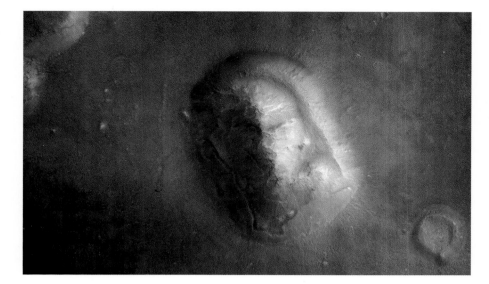

The HiRISE instrument on Mars Reconnaissance Orbiter (MRO) captured this image in 2007 of an eroded mesa made famous as the "Face on Mars." The Viking Orbiter image had much lower spatial resolution and a different lighting geometry. *NASA/JPL*

unknown if living creatures ever existed there. Only time and more research will tell if these findings will prove out. If they do, then human opportunities for colonization of Mars expand exponentially. With water, either in its liquid or solid form, humans might be able to make many other compounds necessary to live and work on Mars.

The Viking landings on Mars represented a series of significant strides forward in planetary exploration. First, it was the first successful landing on the red planet. While soft landings on the Moon had taken place in the 1960s, and on Venus in the early 1970s, not until 1976 did any mission make a soft landing on the most Earth-like planet and complete a mission. Second, through data collected by scientific instruments on these landers, scientists demonstrated the stark and hostile nature of the environment on the surface of Mars. Finally, the Viking mission dashed the rather simple hopes and dreams about the potential for life at other places of the solar system, replacing it with more sophisticated understanding. Knowledge gained in this mission changed the nature of the discussion and the investigation about the central question of space science: are we alone in the universe?

The artifact in the National Air and Space Museum's collection is a structural dynamics test article transferred from NASA in 1979. While Vikings 1 and 2 were on Mars, this third vehicle was used on Earth to simulate their behavior and to test their responses to radio commands. Earlier, it had been used to demonstrate that the landers could survive the stresses they would encounter during the mission. On January 7, 1981, NASA formally transferred ownership of the Viking 1 lander on Mars to the National Air and Space Museum of the Smithsonian Institution. That lander is virtually identical to the "proof test article" displayed in the Museum. NASA Administrator Robert A. Frosch renamed the Viking 1 lander on Mars the Thomas A. Mutch Memorial Station, honoring NASA's fourth associate administrator for space science and the former leader of the Viking Lander Imaging Science Team. Mutch had disappeared on October 6, 1980, while climbing in the Himalayas.

—*Roger D. Launius*

VIKING LANDER
SPECIFICATIONS

MANUFACTURER: Martin Marietta
LENGTH: 10 ft. (3m)
HEIGHT: 6 ft. 6 in. (2m)
WIDTH: 10 ft. (3m)
WEIGHT, UNFUELED: 1,270 lbs. (576 kg)
LAUNCH VEHICLE: Titan III E-Centaur

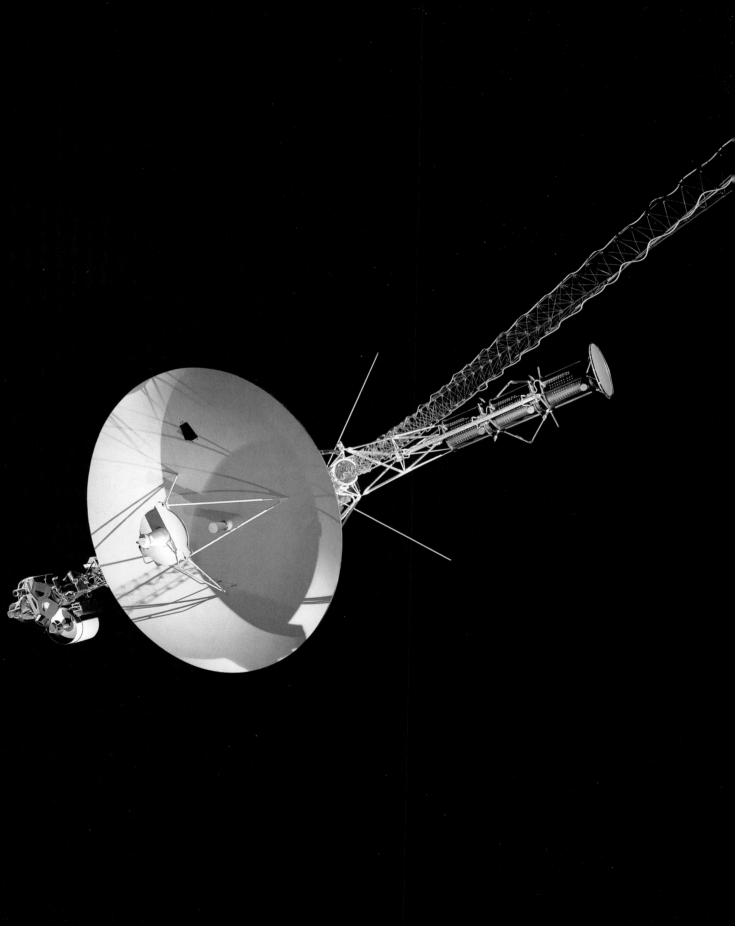

VOYAGER CHIEF SCIENTIST Edward C. Stone once called it "the little spacecraft that could." That may have been an understatement. Voyagers 1 and 2 have accomplished what planetary scientists had only dreamed of in the 1960s: a bold "Grand Tour" of the outer solar system in which all of the four gas giants—Jupiter, Saturn, Uranus, and Neptune—could be visited during a single mission. In the process, these two spacecraft revolutionized knowledge of the gas giants of the outer solar system, their icy moons, and the heliopause, where the solar system meets interstellar space.

The Voyager project emerged from the minds of planetary scientists in the 1960s. Gary A. Flandro, Michael Minovitch, and other scientists and engineers at the NASA Jet Propulsion Laboratory had discovered that once every 176 years, the giant planets on the outer reaches of the solar system all gather on one side of the Sun, and such a configuration was due to occur in the late 1970s. It would be a tragedy, planetary scientists argued, not to take advantage of this opportunity. To do so, of course, would require a drawn-out planning, development, and construction process followed by a lengthy operational period. The Voyagers have so far spent forty years on their journey to the edge of the solar system and beyond.

This short-lived planetary lineup offered a crucial advantage: As the spacecraft passed each planet, gravity would bend its flight path and increase its velocity enough to deliver it to the next destination. Known as "gravity assist," this complex process

VOYAGERS 1 AND 2 // 9

gave the spacecraft a "slingshot" boost at each planet. Neptune, the outermost planet in the mission, thus could be reached in twelve rather than thirty years. Even tiny Pluto, then counted as the ninth planet, could be reached if there was more than one spacecraft, by using Jupiter and Saturn gravity assists to send one in that direction.

In 1964, NASA proposed Pioneers 10 and 11 as the first probes to explore Jupiter and Saturn, missions that can be viewed as a precursor to the Grand Tour. This heady decision would open the outer solar system to American exploration as part of the Cold War space race. Severe budgetary and technical constraints hampered those projects, however, delaying their launch until 1972 and 1973, respectively. Once sent on their way, however, Pioneer 10 and 11 yielded invaluable scientific data. Designed to last for thirty months, they performed for more than twenty years, returning revolutionary scientific knowledge of the two largest gas giants.

NASA declared in 1990 that Pioneer 11 officially departed the solar system by passing beyond the Kuiper Belt and headed into interstellar space toward the center

The Voyager spacecraft on display at the National Air and Space Museum is a backup of the original twin spacecraft sent on a "Grand Tour" of the outer solar system. *NASM*

Voyager 1 and Voyager 2 were the third and fourth human artifacts to be launched into interstellar space. Pioneers 10 and 11 had carried small metal plaques identifying their place of origin for the benefit of any extraterrestrials who might encounter them in the distant future. For the Voyager missions, Cornell University astrophysicist and space popularizer Carl Sagan wanted to build on this earlier Pioneer 10/11 model, in which he had been the driving force, by proposing to affix a type of "time capsule" to the exterior of the two Voyager spacecraft in the form of an analogue recording. As he said, "The spacecraft will be encountered and the record played only if there are advanced spacefaring civilizations in interstellar space. But the launching of this 'bottle' into the cosmic 'ocean' says something very hopeful about life on this planet."

The Voyager "Sounds of Earth" record proved to be one of the most popular ways of engaging the public with the mission. Even U.S. President Jimmy Carter recorded greetings to this life beyond in 1977, evocatively commenting, "This is a present from a small, distant world, a token of our sounds, our science, our images, our music, our thoughts, and our feelings. We are attempting to survive our time so we may live into yours."

Both Voyagers 1 and 2 included a gold-plated aluminum cover enclosing the "Sounds of Earth," which included etchings providing instructions for playing the record. The cover was secured in plain view on the outside of the two Voyager probes, complete with the stylus cartridge required to play the disc. Sagan believed that barring a major collision with something else, the spacecraft and record should last for as much as a billion years.

Inside the case, the twelve-inch, gold-plated disc had two copper sides bonded back-to-back. The side that faced the spacecraft contained all of the images, as well as human greetings, various sounds of Earth, and a third of the music selections. The outer side

LEFT: This gold aluminum cover was designed to protect the Voyager record from micrometeorite bombardment, yet it also provides the finder a key to playing it. The explanatory diagram appears on both the inner and outer surfaces of the cover, as the outer diagram will be eroded in time. *NASA*

RIGHT: These gold-plated copper discs aboard Voyagers 1 and 2 contain greetings in sixty languages, samples of music from different cultures and eras, and natural and human-made sounds from Earth, plus electronic information that an advanced civilization could convert into diagrams and photographs. *NASA*

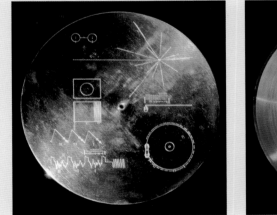

A record is mounted on Voyager 1 prior to launch. *NASA*

consisted entirely of music. Sagan, who led the committee that decided what to include, chose sounds and images that portrayed the diversity of life and culture on Earth. Assembling 116 images and a variety of natural sounds, they also added spoken greetings from Earth in fifty-five languages. The spoken greetings included Akkadian, spoken in Sumer about six thousand years ago, and Wu, a modern Chinese dialect, as well as many current languages. There was also an eclectic, ninety-minute musical program, in which Chuck Berry's "Johnny B. Goode" and Mozart's "Magic Flute" vied with a Zairian Pygmy girls' initiation song, a shakuhachi piece from Japan, and "Dark Was the Night," written and performed by Blind Willie Johnson.

The Voyager record captured the imagination of Americans. Speculation abounded about the possibilities of contact with life beyond this planet. In a 1977 episode of the late-night comedy-variety show *Saturday Night Live*, a segment with a psychic named Cocuwa—played by comedian Steve Martin—predicted that the first message to come back to Earth after sending out the record beyond would be four simple words: "Send more Chuck Berry."

The Voyager record has been referenced in many science fiction films and television series, and sometimes was a major part of the plot. In *Star Trek: The Motion Picture* (1979), a massive spacecraft known as *V'ger* menaces Earth. As the plot unfolds, the crew of the USS *Enterprise* defeats *V'ger*, which turns out to be built around a damaged Voyager seeking its origins from the record. In addition, *Starman* (1984) featured an extraterrestrial visiting Earth, invited by the Voyager record. The potential for alien contact with Voyager is remote, of course, but remains a tantalizing possibility.

The Voyager record in the collection of the National Air and Space Museum was one of several created before the launch. NASA transferred it to the Smithsonian's collections in 1978.

of the Milky Way galaxy. This claim may have been a bit premature, because later scientific findings from Voyager 1 have suggested that it actually departed the heliosphere. These are fine points for most people, and which was first is really very much a question of how one defines the outer limit of the solar system. If the Oort Cloud of comets is included, then none of them will leave the region of solar influence for a very long time. Regardless, there is a general perception that Pioneer 11 was the first human-made object to leave the solar system.

Pioneer 11 ended its mission on September 30, 1995, when NASA received its last transmission. Pioneer 10 lasted even longer. Ground stations received its last, very weak signal on January 22, 2003. At last contact, Pioneer 10 was 7.6 billion miles from Earth, or eighty-two times the nominal distance between the Sun and the Earth. At that distance, it takes more than eleven hours and twenty minutes for the radio signal, traveling at the speed of light, to reach the Earth. Pioneer 10 will continue to coast silently as a ghost ship into interstellar space, heading generally

A prototype Voyager spacecraft at NASA's Jet Propulsion Laboratory is subjected to vibration tests simulating the expected launch environment. The large parabolic antenna at the top is 12.1 feet (3.7 meters) in diameter and is used for communicating with Earth over the great distances from the outer planets. The spacecraft received electrical power from three nuclear power sources (lower left). The shiny cylinder on the left side under the antenna contained a folded boom, which extended after launch to hold a magnetometer instrument thirteen meters away from the spacecraft body. The truss-like structure on the right side is the stowed instrument boom that supported three science instruments and a scan platform, which allowed the accurate pointing of two cameras and three other science instruments. *NASA*

for the red star Aldebaran, which forms the eye of the constellation Taurus (The Bull). Aldebaran is about sixty-eight light years away, and it will take Pioneer 10 more than two million years to reach it. "From Ames Research Center and the Pioneer Project, we send our thanks to the many people at the Deep Space Network (DSN) and the Jet Propulsion Laboratory (JPL), who made it possible to hear the spacecraft signal for this long," said Pioneer 10 Flight Director David Lozier at the time of the last contact.

As remarkable as Pioneer 10 and Pioneer 11 were, they were only preliminaries for the stunning harvest of scientific data that came from the flights of Voyagers 1 and 2. This effort was an order of magnitude more complex than the Pioneer twins. As early as 1966, leaders at NASA's Jet Propulsion Laboratory had proposed a Grand Tour of the four outer gas giants. That mission would require a complicated, self-repairing spacecraft that would last much longer than ones used for the inner planets. NASA formally proposed the Grand Tour in 1971, but canceled it in 1972 due to budget shortfalls and infighting among scientists over its expense

Voyager 2 was launched on August 20, 1977, aboard a Titan IIIE-Centaur rocket, sixteen days before its partner. *NASA*

and complexity. It then reemerged in 1973 as Mariner-Jupiter-Saturn 1977 (MJS 77), rather than the four-planet Grand Tour. Undeterred, mission designers opted for a robust spacecraft that they hoped would fly through the outer reaches of the solar system, enduring for more than a quarter century while reliably operating a range of scientific instruments.

To simplify the mission and limit the expense, the Voyager planners focused on intensive flyby studies only of Jupiter and Saturn, but with an $865 million budget in the 1970s (less than a third of the cost of a comparable mission in the early twenty-first century), engineers designed the two Voyagers to conduct as much science and to last as long as possible. Voyager 2 flew first, lifting off on August 20, 1977, from the Kennedy Space Center. Voyager 1 followed on September 5, 1977, taking a faster trajectory. It soon overtook its sibling and reached Jupiter and Saturn first—hence, NASA's decision on numbering.

The twin Voyagers carried a complex array of scientific instruments to provide measurements and imagery of the outer solar system. Two long antennas allowed for radio astronomy and plasma wave studies, while magnetometers on a thirteen-meter boom offered a wide spectrum of measurements of magnetic fields. Instruments to detect and measure cosmic rays and charged particles were on the spacecraft, while spectrometers operating in the infrared and ultraviolet zones collected data

This picture of the Earth and Moon in a single frame, the first of its kind ever taken, was recorded on September 18, 1977, by Voyager 1 7.25 million miles (11.66 million kilometers) from Earth. Visible is eastern Asia, the western Pacific Ocean, and part of the Arctic. Voyager 1 was directly above Mt. Everest on the night side of the planet when the picture was taken. *NASA*

across a broad range. A suite of visible-light cameras completed the instrumentation, including two television-type cameras—one low-resolution and the other high-resolution—each with eight filters serving vidicons of the type used in earlier space missions. These instruments proved exceptionally sturdy and worked effectively throughout the mission.

Communications, command, and control systems enabled controllers on Earth to guide the spacecraft and investigate the outer solar system, as well as to communicate data back to the home planet. The Attitude and Articulation Control Subsystem (AACS) controlled spacecraft orientation, while propulsion rockets enabled course corrections. For on-board electrical power, both Voyagers used radioisotope thermoelectric generators (RTGs). The RTG is a critical technology for these outer planet missions. For spacecraft traveling for long periods in deep space far from the energy

Jupiter and its four planet-sized moons, called the Galilean satellites, were photographed in early March 1979 by Voyager 1. In this collage, they are not to scale but are in order of their distance from Jupiter: reddish Io (upper left) and then Europa (center), Ganymede, and Callisto (lower right). Many other much smaller satellites circle Jupiter, several inside Io's orbit and others millions of miles from the planet. *NASA*

of the Sun's rays, RTGs are almost the only way to satisfy mission requirements. It is a very simple technology: Radioactive plutonium 238 decays, producing heat that flows through thermocouples to the heat sink, generating electricity in the process. The thermocouples are connected through a closed loop that feeds electrical current to the power management system of the spacecraft. This method provided a very stable, long-lived power source for Voyagers 1 and 2.

Voyager 1 began observations of Jupiter in January 1979, with its closest approach to the planet on March 5. During its encounter with Jupiter, Voyager 1 took almost nineteen thousand pictures and made many other scientific measurements. Voyager 2 followed between April and August 1979 with its reconnaissance of the Jovian system, making its closest approach on July 9. Collectively, they took more than thirty-three thousand pictures of Jupiter and its five major satellites. Those

EDWARD C. STONE

No one has been more closely identified with the flights of Voyager 1 and Voyager 2 than Edward Stone, the chief scientist of the program and still a key part of the scientific team as the spacecraft leave the solar system. Stone, born in Knoxville, Iowa, on January 23, 1936, had become involved in space science after joining the physics department of the California Institute of Technology in Pasadena, California, in 1967. Then focused on cosmic radiation, he flew experiments on the United States' classified DISCOVERER (or CORONA) series satellites throughout the 1960s and on longer duration missions into the 1970s. This research led to a longitudinal set of scientific data about radiation in near space.

Stone joined what would become the Voyager program in 1972, when the Jet Propulsion Laboratory's Bud Schurmeier, the first project manager for the Mariner-Jupiter-Saturn 1977 mission, invited him to participate as the mission's lead scientist. Previously, Stone had proposed a cosmic ray experiment for both Voyagers; becoming project scientist allowed him to shape the missions' scientific endeavor and to extend his own scientific investigations. Stone enjoyed a much larger role as coordinator for all eleven scientific teams, and he also served as the central liaison with the spacecraft engineering team. He started part-time but soon transitioned full-time to JPL, which Caltech ran for NASA, because of the breadth of his Voyager mission responsibilities. He maintained his faculty position, however, and eventually returned to the university.

While Stone was chief scientist, Voyagers 1 and 2 undertook their spectacular encounters with Jupiter and Saturn, Voyager 2 went on to Uranus and Neptune, and both Voyagers have continued their investigations into deep space. Stone has demonstrated deftness in balancing the needs and expectations of mission scientists and project engineers, negotiating agreements on a range of delicate matters. He also proved remarkably successful at dealing with the political establishment, explaining the mission to those who had little understanding of its intricacies, and helping to maintain a coalition of supporters across a broad spectrum of officialdom.

first images and data returned from the solar system's largest planet were visually stunning. While the clouds and storms of Jupiter, like the Great Red Spot, had been observed for centuries by astronomers, the scientific knowledge gained through close-up inspection proved path-breaking. The data from the Galilean satellites demonstrated how much there was yet to learn. The Voyagers found active volcanism on the satellite Io, the greatest surprise of the exploration because no other bodies in the solar system had at that point been found to have active volcanoes. Moreover, the cracked ice shell of Europa, which has almost no craters, raised for the first time

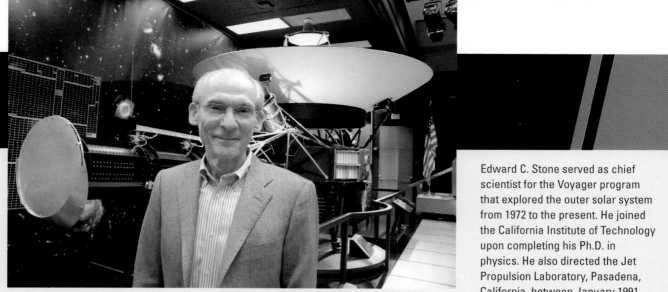

Stone was an obvious and politically astute choice to succeed Gen. Lew Allen as JPL director when Allen retired in 1990. Stone served for a decade, during a time when NASA was under the gun from three presidential administrations to reduce budgets—although few of those officials wanted similar reductions in either the quality or quantity of space science missions. Under his tenure, planetary science enjoyed incredible successes and crushing disasters. Other programs included the Galileo mission, which undertook a lengthy exploration of Jupiter and its moons, and the Cassini/Huygens spacecraft to explore the Saturnian system, launched in 1997. He also presided over the loss of Mars Observer in 1993, which nearly led to the termination of Mars exploration, as well as its resurrection on a more modest budget in the latter part of the decade.

In 1999, Stone and his advisors lost two Mars missions: Mars Climate Orbiter and Mars Polar Lander, harsh reminders that squeezing cost, schedule, and reliability—the "iron triangle" of project management—could take an organization only so far. A measure of restructuring followed, with a Flight Projects Directorate that codified successful project practices to ensure that these accidents did not happen again.

In April 2001, Ed Stone retired as director of JPL and returned to Caltech's physics department. He served as vice provost remained engaged in a range of scientific investigations, including as project scientist on the Voyager Interstellar Mission into the second decade of the twenty-first century.

the possibility of a subsurface liquid ocean and the potential for aquatic life there. Finally, the changes in charged particles between the Pioneer 10 /11 missions and Voyager 1/2 missions established that Jupiter's magnetosphere was dynamic and constantly changing.

The Voyager 1 and 2 encounters with Jupiter were superb successes. Although astronomers had studied Jupiter through telescopes on Earth for centuries, scientists were surprised by many of the Voyagers' findings. They revealed for the first time that the Great Red Spot was a complex storm moving in a counterclockwise

This Voyager 2 image of Jupiter extends from the equator to the southern polar latitudes. A white oval, different from the one observed in a similar position at the time of the Voyager 1 encounter, is situated south of the Great Red Spot. A region of white clouds now extends east of and around its northern boundary, preventing small cloud vortices from circling the feature. The disturbed region west of the red spot had also changed. It shows more small-scale structure and cloud vortices. The picture was taken on July 3, 1980, from 3.72 million miles (6 million kilometers). *NASA*

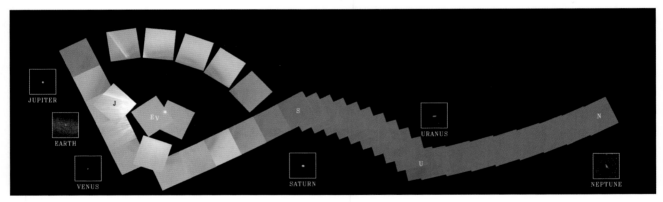

The cameras of Voyager 1 on February 14, 1990, pointed back toward the Sun and took a series of pictures of the Sun and the planets, making the first ever "portrait" of our solar system as seen from the outside. In the course of taking this mosaic consisting of a total of sixty frames, Voyager 1 made several images of the inner solar system from a distance of approximately 4 billion miles and about 32° above the plane of the system. From Voyager's great distance, Earth and Venus are mere points of light, less than the size of a picture element even in the narrow-angle camera. Coincidentally, Earth lies right in the center of one of the scattered light rays resulting from taking the image so close to the Sun. Scientist Carl Sagan nicknamed this close-up of our home planet the "Pale Blue Dot." *NASA*

direction. Together, the two observed the eruption of nine volcanoes on Io, and there is evidence that other eruptions occurred between the Voyager encounters. Plumes from Io's volcanoes extended to more than 190 miles above the surface; these plumes resulted from heating through tidal interactions with Jupiter and other moons.

Both Voyagers 1 and 2 then moved on to Saturn, reaching the planet nine months apart, in November 1980 and August 1981. These Saturn encounters offered extended, close-range observations with visual imagery and through high-resolution

This montage of the Saturnian system was prepared from Voyager 1 images taken during its encounter in November 1980. Dione is in the forefront, Saturn rising behind, Tethys and Mimas fading in the distance to the right, Enceladus and Rhea off Saturn's rings to the left, and Titan in its distant orbit at the top. *NASA*

measurements. Among other findings, the Voyagers discovered that Saturn's atmosphere is almost entirely hydrogen and helium. The complex moon system at Saturn also proved fascinating for scientists. Overall, the spacecraft found five new moons and a ring system consisting of thousands of bands, discovered a new ring (the G-ring), and found "shepherding" satellites on either side of the F-ring that kept it well defined.

A major objective of the Saturn encounters was to image Titan, the largest moon and one of the most interesting bodies in the solar system because it is the only satellite with a substantial atmosphere. After encountering Jupiter, Voyager 1 went to within 2,500 miles (4,000 km) of the moon while Voyager 2 followed a less Titan-oriented trajectory. When flight controllers opted for a close approach, it ensured that the spacecraft would be unable to continue on to Uranus or Pluto, thus ending Voyager 1's Grand Tour possibilities because it sent the spacecraft flying out of the solar system's ecliptic plane. But Voyager 1's encounter with this inviting moon revealed Titan as a

At the dawn of the twenty-first century, both Voyagers continued to provide important scientific data about the heliosheath and heliopause, where the flow of the solar wind eventually stops as it ploughs into the charged particles and atoms between the stars, which are embedded in the magnetic field of our galaxy. NASA has called this investigation the Voyager Interstellar Mission. While imagery is no longer possible from the Voyagers due to declining electrical output from the radioisotope thermal generators, scientists have been able to use low-powered instruments to continue study of the outer reaches of the solar system. Energetic particle, cosmic ray, plasma wave, and magnetic-field experiments aboard the spacecraft have recently provided new data on the nature of this completely unknown region.

Voyager 1 is not only currently the farthest spacecraft from Earth, it is also traveling toward the head of the tear-shaped bubble that the Sun blows in the ultra-thin interstellar medium as our solar system travels through space. It is thus the first human object that is likely to encounter the interstellar realm. On August 25, 2012, based on the level of low-energy particles encountered and the cosmic rays measured and a change in the density of the surrounding charged particles, Voyager 1 completed passage through the heliopause and became the first human object to enter the interstellar medium. But it took a year before mission scientists could confirm that fact.

Based on preliminary data, on December 3, 2012, Voyager project scientist Edward C. Stone and his colleagues Stamatios Krimigis and Leonard Burlaga stated in a NASA press conference: "Voyager has discovered a new region of the heliosphere that we had not realized was there. We're still inside, apparently. But the magnetic field now is connected to the outside. So it's like a highway letting particles in and out." In essence, according to Stone and colleagues, the pressure of the interstellar medium compresses the Sun's field. This creates a local magnetic field that is about ten times more powerful than that encountered previously.

In April and May 2013, new evidence came to light from the plasma wave experiment on Voyager 1, which began detecting oscillations in the charged particles surrounding the space-craft. These waves were caused by gases from a particular strong set of solar storms pushing against the boundary between the region of solar influence and the interstellar medium. The

place of complexity, having thick clouds and water ice. The spacecraft also found that Titan's atmosphere was composed of 90 percent nitrogen—also the major constituent of Earth's atmosphere—and that the pressure and temperature at the surface was 1.6 atmospheres and −180°C. Its uniqueness and its mystery—a global haze obscured the surface, making photos impossible—made it a place that scientists wanted to learn more about. That came to pass at the dawn of the twenty-first century with the

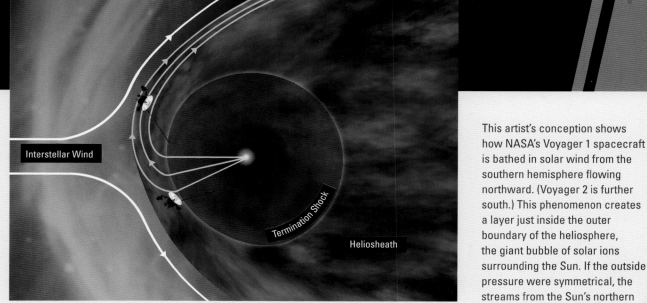

Interstellar Wind

Termination Shock

Heliosheath

This artist's conception shows how NASA's Voyager 1 spacecraft is bathed in solar wind from the southern hemisphere flowing northward. (Voyager 2 is further south.) This phenomenon creates a layer just inside the outer boundary of the heliosphere, the giant bubble of solar ions surrounding the Sun. If the outside pressure were symmetrical, the streams from the Sun's northern hemisphere above the plane of the planets would all turn northward and the streams from the southern hemisphere would all turn southward. However, the interstellar magnetic field presses more strongly on the boundary in the southern hemisphere, forcing some of the solar wind from the south to be deflected north toward Voyager 1. *NASA*

experiment's scientists, led by Donald Gurnett and William Kurth, then found weak waves in their data from October and November 2012, due to an earlier set of solar storms. Extrapolating backwards from these two incidents, they concluded that the spacecraft had indeed entered interstellar space on August 25, 2012. The Voyager science team met three times and agreed after some debate, although the transition between the two regions was more complex and rather different than expected. One thing is clear: Voyager had accomplished an historic milestone in the human exploration of space.

Most of spacecraft systems on the two Voyagers have been shut down already to save power, and even the interstellar mission is due for termination in the next few years. NASA has announced that it will probably shut down the gyroscopes for Voyager 2 in 2015 and do the same in 2016 for Voyager 1. The spacecraft can continue to point to Earth, but the gyros are needed if the orientation of the spacecraft is to be changed, which is a key to making important measurements. But at least until that time, the two spacecraft can continue to send back data to help us understand how the region of solar influence interacts with the material between the stars. And Voyager 2 may make the same transition into interstellar space as it heads outwards in a somewhat different direction. More than three decades after their launch, the Voyagers continue their historic missions.

Cassini/Huygens mission. In addition to Titan, Voyager 1 also photographed the moons Mimas, Enceladus, Tethys, Dione, and Rhea. Voyager 1's last imaging sequence was a portrait of most of the solar system, showing Earth and six other planets as sparks in a dark sky lit by a single bright star, the Sun.

With the success of Voyager 1 at Titan, the spacecraft's mission objective was achieved; Voyager 2 thus approached the Saturnian system on a trajectory that

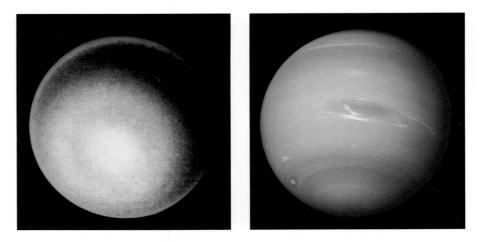

LEFT: This computer enhancement of a Voyager 2 image emphasizes the high-level haze in Uranus's upper atmosphere. The underlying clouds are obscured. *NASA*

RIGHT: This picture of Neptune was taken by Voyager 2's narrow angle camera, 4.4 million miles (7 million kilometers) from the planet, and four days twenty hours before closest approach. Visible are the Great Dark Spot and its companion bright smudge; on the west limb are the fast moving bright feature called Scooter and the little dark spot. North of these, a bright cloud band similar to the south polar streak may be seen. Years later, when the Hubble telescope imaged the planet, these atmospheric features had changed, indicating that Neptune's atmosphere is dynamic. *NASA*

allowed it to continue on to Uranus and Neptune. It came within 63,000 miles (101,000 kilometers) of Saturn on August 26, 1981. Stunning images of the rings, a newly discovered F-ring, and its shepherding moons proved exciting for all. It also imaged the moons Hyperion, Enceladus, Tethys, and Phoebe.

Flybys of the two outermost giant planets, Uranus and Neptune, proved irresistible to mission scientists for Voyager 2. This necessitated a conscious decision at NASA to determine whether or not to send Voyager 2 on to Uranus or expand the investigation of Titan and the rings of Saturn. The two activities were mutually exclusive. As Voyager project manager Raymond L. Heacock put it in October 1980: "Voyager 2 health and propellant supply have preserved the opportunity to continue past Saturn to the remote Uranus, and possibly even Neptune, planetary domains. This is an opportunity that may not occur again until after the turn of the century. For this reason, and because of the paucity of data about the Uranus and Neptune systems, there is strong support from the scientific community to continue to Uranus. We have also assessed our ability to operate Voyager 2 at these greater distances and have not uncovered any major problems." NASA officials agreed. As Voyager 2 flew across the solar system, scientists reprogrammed it via radio to aim for the last two gas giants.

In January 1986, Voyager 2 reached distant Uranus, the seventh planet from the Sun, and at its closest approach, January 24, the spacecraft came within 50,600 miles (81,500 kilometers) of Uranus's cloud tops. It returned thousands of images and other types of measurements about the planet, its moons, and its powerful magnetosphere. Most important, Voyager 2's instruments revealed two new rings that had not been detected from Earth as well as ten previously unseen moons, adding to the five large satellites already known.

From Uranus, Voyager 2 journeyed on to Neptune and reached the planet in August 1989. It flew within 3,000 miles (4,950 kilometers) of Neptune's north pole, and on to Triton, its largest moon, the last solid body the spacecraft encountered as it sailed outward toward the end of the solar system.

Voyager 2 explored all four of the giant outer planets, including their unique systems of rings and magnetic fields, as well as forty-eight of their moons. The Grand

Tour was successfully completed in August 1989 with the Neptune flyby. Overall, these two spacecraft returned information to Earth that revolutionized planetary science, helping resolve some key questions while raising intriguing new ones about the origin and evolution of the solar system. The two Voyagers took well more than one hundred thousand images of the outer planets, rings, and satellites as well as millions of chemical spectra and magnetic and radiation measurements. They discovered rings around Jupiter, shepherding satellites in Saturn's rings, new moons around Uranus and Neptune, and geysers on Triton. On November 17, 1998, Voyager 1 overtook Pioneer 10 as the most distant human-made object from Earth, reaching a distance of 69.419 astronomical units (AU), and it has continued outward ever since in the general direction of the star Gliese 445 in the constellation Camelopardalis.

A list of major scientific discoveries from the Voyager missions to the outer planets included the following, as reported by the National Space Science Data Center:

- The Uranian and Neptunian magnetospheres, both of them highly inclined and offset from the planets' rotational axes, suggesting their sources are significantly different from other magnetospheres.
- The Voyagers found twenty-two new satellites: three at Jupiter, three at Saturn, ten at Uranus, and six at Neptune.
- Io has active volcanism, the only solar system body other than the Earth to be so confirmed.
- Triton has active geyser-like structures and an atmosphere.
- Jupiter, Saturn, and Neptune all had auroral zones at their poles.
- Jupiter has rings. Saturn's rings were found to contain spokes in the B-ring and a braided structure in the F-ring. Two new rings were discovered at Uranus, and Neptune's rings, originally thought to be only ring arcs, were found to be complete, albeit composed of fine material.
- Neptune, originally thought to be too cold to support atmospheric disturbances, has large-scale storms (notably the Great Dark Spot).
- While the primary mission ended in 1989, the Voyager Interstellar Mission (VIM) has continued to the present.

The National Air and Space Museum has on display a stunning model of the Voyager spacecraft. It was created using the Development Test Model (DTM) with additional parts attached to make it look like the flight Voyagers. Manufactured by the Jet Propulsion Laboratory (JPL) in Pasadena, California, it was acquired by the National Air and Space Museum in 1977 and placed on display in the *Exploring the Planets* exhibition shortly thereafter. Interestingly, this Voyager model made a real contribution to space exploration. In 1987, JPL removed the DTM bus for use in developing the Magellan Venus spacecraft, which was similar in design, and replaced it with a replica.

—*Roger D. Launius*

Voyager 1 acquired this image of Io on March 4, 1979, from 304,000 miles (490,000 kilometers) away. An enormous volcanic explosion can be seen silhouetted against space over Io's bright limb. The brightness of the plume has been increased as it is normally extremely faint, whereas the relative color of the plume (greenish white) has been preserved. Solid material had been thrown up to an altitude of about 100 miles (160 kilometers), requiring an ejection velocity of about 1,200 mph (1900 km/h). The vent area is a complex circular structure consisting of a bright ring about 300 kilometers in diameter and a central region of irregular dark and light patterns. *NASA*

VOYAGER SPECIFICATIONS

LENGTH: 57 ft. (17.37m)
HEIGHT: 9 ft. 6 in. (2.90m)
WIDTH: 21 ft. (6.40m)
HIGH-GAIN ANTENNA:
  12 ft. diameter (3.66m)
WEIGHT, UNFUELED:
  1,800 lbs. (816.5 kg)
MANUFACTURER: NASA/Jet
  Propulsion Laboratory
LAUNCH VEHICLE:
  Titan III E-Centaur

TWICE *DISCOVERY* STOOD POISED on the launch pad to take American astronauts into orbit after a long pause in the Space Shuttle program. With the loss of *Challenger* during its tenth launch in 1986 and *Columbia* during its twenty-fifth reentry in 2003, *Discovery* became the return-to-flight vehicle. In fact, it bore that responsibility three times; the first two post-*Columbia* missions, both on *Discovery*, were test flights to validate remedies for the causes of that fatal accident. Executing their missions flawlessly, three *Discovery* crews and their spacecraft restored confidence, albeit chastened, in the Space Shuttle program.

Discovery was the champion of the shuttle fleet, and not simply because it returned astronauts to space and flew more missions—thirty-nine in all—than did *Columbia*, *Challenger*, *Atlantis*, and *Endeavour*. Like a champion, it served longer (twenty-seven years) and spent more time at its peak (a cumulative 365 days in space). Its flight history began in 1984 as the fleet was starting its busiest two years, and it ended in 2011 as the program wound down. *Discovery* flew every type of mission and served every purpose the Space Shuttle was meant to serve. *Discovery* had no rival in the variety of its missions and the range of firsts it attained.

Discovery's story is a digest of the Space Shuttle story as a whole. Its thirty-nine-episode narrative rises with the successes and falls with the disappointments that mark the four-decade effort by the United States to make human spaceflight in Earth orbit routine, practical, economical, and safe. *Discovery* is a sturdy icon for the shuttle era.

SPACE SHUTTLE
DISCOVERY
/10

The Space Shuttle came into being in the 1970s as the answer to two questions: "What should the United States do in space after sending men to the Moon?" and "How can the United States continue putting people in space at a lower cost than the Moon missions?" With no national appetite for an expensive grand venture—a space station or a mission to Mars—a fragile consensus settled on a new space transportation system, a shuttle for intermittent missions around Earth, that was supposed to reduce the cost of spaceflight. If it proved successful, it might later pave the way to a space station or to deep-space expeditions.

The central element, often called the workhorse or space truck or spaceplane, was a reusable orbiter, large enough to carry both people and payloads and versatile enough to keep dreams alive for a more exotic future. Attached to twin solid rocket boosters and pumping liquid propellants from an enormous external tank into its three internal launch engines, the vehicle streaked from Earth to orbit in eight and a half minutes. Having shed the boosters and tank during ascent, the spacecraft operated

Discovery launched on the first of its thirty-nine missions on August 30, 1984, with a six-person crew ready to deploy three communications satellites and conduct science experiments. *NASA*

in the altitude range of 115 to 400 miles (185 to 644 kilometers) on stays ranging from two to eighteen days. Covered with a novel thermal protection system made of ceramic tiles and blankets, the orbiter descended from space to land on a runway and then moved into a servicing bay to be readied for its next mission.

The orbiter resembled an airplane more than a spacecraft. Unlike space capsules, it had wings and wheels and was as large as a Boeing 737 passenger jet airliner. Its wide delta wings enhanced maneuverability for its unpowered glide back to land. It had a tall tail fin, or vertical stabilizer, with a rudder/speed brake, at the base of which were two bulbous pods that housed orbital maneuvering engines and aft reaction control system thrusters, all used primarily in space. The crew cabin had two levels: an upper flight deck plus middeck living quarters that accommodated up to seven (and twice eight) people with comforts comparable to camping or houseboating. Behind the crew cabin stretched a sixty-foot by fifteen-foot (18.3-meter by 4.6-meter) payload bay under long doors that remained open in orbit. This was the cargo container for satellites, laboratories and observatories, large experiments, and, later on, the structural beams and modules of the International Space Station.

The reusable Space Shuttle was conceived to reduce the astronomical costs of human spaceflight and to begin using space near Earth for practical purposes—to begin living and working in space on a routine basis. In the heady early days of development, planners envisioned spaceflight service as regular as an airline, with a fleet of five or more orbiters launching from sites in Florida and California as often as once a week. This proved overly optimistic for a variety of reasons, not the least of which was that the shuttle itself was a product of compromise. Rather than completely reusable, it was only partially so, a measure that kept its development cost low. However, that compromise drove its operational costs higher. The shuttle was the most technically complex space vehicle ever built, and like a high-performance racecar it required constant attention, whether on the ground or in space. NASA had to maintain a standing army of engineers and technicians to maintain the shuttles. Approval never came for a five-vehicle fleet, and the western site was eventually canceled without ever being used for a shuttle launch.

The space transportation system was designed to serve all of the nation's launch needs for commercial, scientific, and national security access to space. The plan called for the shuttle to become the sole launch vehicle for all types of payloads. With more onboard engineers and scientists than pilots, shuttle crews offered retrieval of errant satellites, in-orbit servicing of balky or failed equipment, hands-on laboratory research, and the skills for assembly of large space structures. Again, for various reasons, the anticipated demand for shuttle flights did not materialize, and operational costs were not recouped from eager customers.

Skeptics doubted the efficacy of the Space Shuttle before it began service, and they continued to challenge the wisdom of this approach to spaceflight throughout its history. Yet from the successful first launch in 1981 to the 1986 *Challenger* launch tragedy, the shuttle ramped up in frequency and duration of flights. Nine missions launched in 1985. The year 1986 was to have been even busier, with three orbiters and fifteen launches at an average rate of more than one a month. Spaceflight

Discovery made its final touchdown on March 9, 2011, to end the STS-133 mission to the International Space Station. The reusable Space Shuttle orbiter was a combination launch vehicle, crew ship, cargo carrier, and glider. *NASA*

TAKING THE HEAT OF REENTRY

PROTECTIVE TILES AND BLANKETS

Much of *Discovery*'s history is suggested in the discolored mosaic of ceramic tiles and blankets that cover the vehicle. Once pristine white and glossy black, most are now shades of beige and gray, the vast percentage of them dulled to a matte finish from repeated exposure to the searing heat of reentry. Chalk-white streaks, thin and straight as if airbrushed from the corners of the tiles, actually record the angle of the orbiter as it descended through the reentry furnace at twenty-five times the speed of sound—not once, but thirty-nine times.

The orbiter airframe and skin are primarily aluminum alloys and composite materials that must be protected from temperatures above 350°F (175°C). High-speed passage through even the thinnest part of the atmosphere generates temperatures up to 3,000°F (1,650°C). Spacecraft meant to return to Earth are usually fitted with an ablative heat shield that gradually chars away during descent, leaving the shield too damaged for further use. The orbiter demanded a reusable, repairable, and, given the vehicle's almost hundred-ton heft, very lightweight shield.

Discovery's thermal protection system is a carapace of silica tiles and flexible silica blankets that contain more air than solid material. Extremely lightweight, they absorb heat slowly and maintain the vehicle's skin below 350°F (175°C). Black tiles cover the underside, nose, window frames, edges of the stabilizer and wing trailing edges, and areas around all the engine and thruster nozzles from heat up to 2,300°F (1,260°C). White tiles and blankets cover the upper and side surfaces of the wings and fuselage that are exposed to lower temperatures because they are partly protected by the nose-high reentry angle that allows the black underside to bear the brunt of heating. The orbiter's nose and wing leading edges have reinforced carbon-carbon shields to fend off the most extreme heat.

Approximately twenty-four thousand tiles and blankets, each with a unique part number, were custom-fitted in size and shape to their specific location on the orbiter and individually

was beginning to look routine. After the *Challenger* accident brought shuttle flights to a halt for almost three years, the schedule gradually built up to seven and eight missions a year in the 1990s, and launches continued with few pauses for seventeen more years until the second accident, the loss of *Columbia*, temporarily grounded the shuttle again.

Discovery made its debut as the shuttle program was building momentum. Its first mission, STS-41D in August–September 1984, was twelfth in the schedule. Designated as orbiter vehicle OV-103 and delivered to Florida in 1983, *Discovery* was the first production orbiter built after the developmental orbiters *Columbia* and *Challenger*. It was almost seven thousand pounds lighter thanks to structural

installed by hand in a much more laborious process than required for a monolithic heat shield. Yet one virtue of this puzzle-piece shield was that damaged tiles and blankets could be replaced individually without affecting the rest. On average, about a hundred tiles were replaced after each mission. About eighteen thousand original black tiles remain, now streaked and aged to shades of gray in contrast to the shiny black replacements. Despite initial challenges in installing and keeping tiles adhered, this novel solution proved effective and is used on some spacecraft developed for the post-shuttle era.

Discovery has two small but distinctive thermal tile features: extra black tiles resembling teardrops under two window frames, and a short raised wedge under the port wing, the remains of an aerodynamics experiment on its last flights. These features are barely noticeable punctuation marks in *Discovery*'s unique history.

changes in the orbiter's design, enabling it to carry more payload weight. It was immediately brought into service for satellite deliveries. The air force had a stake in *Discovery* for national security missions, having expected this orbiter to be stationed at Vandenberg Air Force Base in California. In its less than two years on duty until the first accident, *Discovery* flew six times, including three consecutive missions, rapidly approaching *Challenger*'s record of nine flights in three years. The future champion was proving its mettle.

Rather than survey *Discovery*'s flight record in strict chronological order, it is more interesting to look at it thematically. The mission log at the end of this chapter presents the chronology for easy reference, but because *Discovery* flew various

EXTRA ARMS FOR ASTRONAUTS

CANADARM REMOTE MANIPULATOR SYSTEM

Seen from the International Space Station on *Discovery*'s last flight, the jointed Canadarm with a straight boom extension rested above the payload bay. All mission crews after the 2003 loss of *Columbia* used this extension and its attached sensor system to inspect the entire vehicle exterior for damage. *NASA*

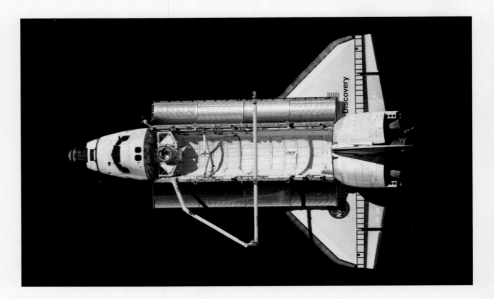

The most important crew aid for spacewalking astronauts, other than the protective extra-vehicular activity (EVA) spacesuit, is the Canadarm, also known as the remote manipulator system (RMS) or robotic arm. This long, jointed arm extends the astronauts' strength and reach. It is used like a crane to move very large items—for example, the Hubble Space Telescope or building blocks of the International Space Station—in and out of the orbiter's payload bay. It also serves as a mobile platform to position astronauts precisely where they need to be while working outside. A Canadarm flew on 90 of the 135 Space Shuttle missions.

This arm (serial number 202) flew on fifteen shuttle missions and four orbiters from 1994 through 2011, including *Discovery*'s last six missions, all to the International Space Station. It is actually a telerobotic arm, remotely operated by an astronaut at the aft flight deck control station inside the orbiter. The arm has shoulder, elbow, and wrist joints plus a snared end effector "hand" for clasping things, and a foot restraint that also attaches to the end. Closed-circuit television cameras mounted on the arm and line-of-sight monitoring through the orbiter's aft and overhead windows assist the operator in controlling its movements.

The Museum chose not to reinstall the Canadarm inside *Discovery*'s payload bay in order to keep it visible as an important icon of the shuttle era. Although it can lift large masses in space, the arm cannot support its own weight on Earth, so it is displayed in a support stand. The Canadian Space Agency has supplied several robotic arms for the shuttle and space station programs.

CANADARM
SPECIFICATIONS

MANUFACTURER:
  SPAR Aerospace
LENGTH: 50 ft. (15m)
DIAMETER: 15 in. (38cm)
WEIGHT: 905 lbs. (410 kg)

types of missions, a review by type reveals more about trends and evolution in the shuttle program. Although every shuttle mission had several objectives, missions generally were designated by their primary purpose or payload as these types: commercial, national security, servicing, science, Mir visits, and International Space Station assembly or logistics. *Discovery* flew multiple missions of each type.

Discovery's first role was to deliver commercial satellites to low Earth orbit, from which they were propelled by attached stages to geosynchronous orbits. Five of *Discovery*'s first six missions served customers from the communications satellite industry, and its first two post-*Challenger* missions deployed NASA Tracking and Data Relay System (TDRS) satellites. Some of the commercial missions also delivered a satellite for the U.S. Navy. For satellite deliveries, the shuttle truly served as a cargo truck; two or three satellites were packed in the payload bay to be released one at a time when the orbiter reached the proper altitude and alignment. *Discovery*'s first mission was the first shuttle flight to carry three satellites. In all, *Discovery* delivered sixteen communications satellites produced by Hughes, AT&T, TRW, and Aerospatiale for the United States, Canada, Mexico, the Arab League, and Australia. They represented the growing market of non-spacefaring nations eager to join a global telecommunications network.

The commercial sector was crucial to the effort to make spaceflight more economical and routine, and NASA's business plan depended on a growing and steady customer base for satellite deliveries and also for research projects. NASA cultivated commercial customers with attractive pricing, launch preparation services, and incentives for repeated flights, including the opportunity for a corporate payload specialist to join the crew. The first commercial payload specialist, Charles D. Walker of McDonnell Douglas, flew on *Discovery* twice and *Atlantis* once to conduct experiments in a manufacturing process with commercial promise. Saudi Arabia and France were able to place payload specialists on *Discovery* to witness the deployment of their satellites, as did Mexico on *Atlantis*. Two members of Congress took advantage of this courtesy and persuaded NASA to put them on crews, Senator Jake Garn on *Discovery* in 1985 and Representative William "Bill" Nelson on *Columbia* in early 1986.

Commercial payloads presented opportunities for another mission type: servicing. Twice *Discovery* crews combined deployments with retrievals for repairs or returns when satellites failed to reach their required

During the 1980s, most *Discovery* missions deployed communications satellites. In this view from the STS-51I mission in 1985, an Australian satellite with attached boost motor rises from the payload bay. Boost stages sent the satellites to geosynchronous orbit about 22,000 miles (35,400 km) away. *NASA*

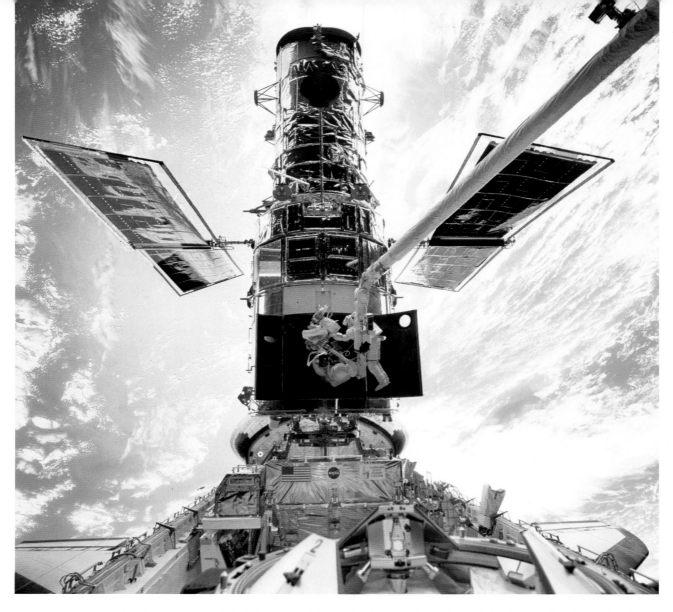

Discovery crews completed two of the five servicing missions that repeatedly extended the life of the Hubble Space Telescope well beyond its planned ten years. In this view from the STS-103 servicing mission in 1999, astronauts Steven L. Smith and John M. Grunsfeld—working at the end of the long robotic arm—have opened a bay to replace gyroscopes in the pointing and attitude control system. *NASA*

orbits. The crew of *Discovery*'s second mission celebrated the first in-orbit satellite retrieval with a "2 Up, 2 Down" sign when they successfully deployed two satellites and picked up two others to bring home for refurbishing. Another *Discovery* crew brought an idle satellite into the payload bay, installed a new ascent motor, redeployed it, and watched it ignite on its intended path. These servicing episodes demonstrated important crew skills and built crew experience for future projects, notably servicing the Hubble Space Telescope and assembling a space station.

Policy changes after the *Challenger* accident took commercial satellites off the shuttle and seriously eroded the commercial market for shuttle flights. Commercial experiments continued to fly as secondary payloads, but *Discovery*'s role soon shifted from satellite delivery to other types of missions.

The Department of Defense reserved *Discovery*'s third flight for the first dedicated, classified, national security shuttle mission, about which little is known beyond the names of the first all-military crew and the first U.S. Air Force payload specialist who was not in the astronaut corps. The primary payload was presumed to be an electronic intelligence satellite. Although the shuttle was designed and developed with national security needs in mind, the air force became reluctant to rely on the shuttle and NASA as its sole launch provider, preferring instead to maintain its own capacity for assured access to space. Even before the *Challenger* accident and subsequent grounding of the fleet, the Department of Defense was easing away from the shuttle. It completed its backlog of planned national security missions when flights resumed and then abandoned the shuttle except for occasional small, unclassified payloads, withdrawing as the big customer NASA had counted on.

From 1984 through 1992, *Discovery* flew four of the ten defense missions. The first two of those were strictly classified, but allegedly spy satellites were deployed for the National Reconnaissance Office. The other two missions were openly linked to the Strategic Defense Initiative ("Star Wars"), one unclassified and the other partially cloaked in secrecy. Editorial cartoonists mocked the secret missions by portraying the shuttle in disguise or invisible, but the point was to question the militarization of space and the place of secrecy in a public space program. The issue dissipated as the air force returned to its preferred rockets and the Space Shuttle moved on to almost exclusively civilian tasks.

In 1990 and again in 1997 and 1999, *Discovery* and its crews made history, first by deploying the heralded Hubble Space Telescope and then by returning twice to service, repair, and redeploy it. *Discovery* was not scheduled for the urgent first visit to install corrective optics, but it drew duty for the second and third of five servicing missions. Spacewalking teams updated the telescope with newer technologies and extended its life by repairing or replacing worn components. With skill and finesse, *Discovery* crews demonstrated the value of humans in space for efficient performance of complex tasks. In-orbit servicing benefited the astronomical community by adding years of continued telescope operations and also built the experience base for future large assembly projects such as the International Space Station.

On missions to the Hubble Space Telescope, the orbiter's name seemed especially apt. Namesake of sailing ships used by explorers Henry Hudson and Captain James Cook, *Discovery* furthered this tradition of exploration and discovery through improved observation of the universe.

Discovery's primary occupation in the 1990s, however, was to support science. On ten missions during the decade, this

Discovery's science missions typically included retrievable research satellites and instruments mounted in the payload bay. This ATLAS 2 suite of instruments for atmospheric and solar physics investigations flew on the STS-56 mission in 1993. *NASA*

Discovery made the first and last of nine shuttle missions to Russia's Mir space station, seen here during the shuttle's final departure in 1998. *NASA*

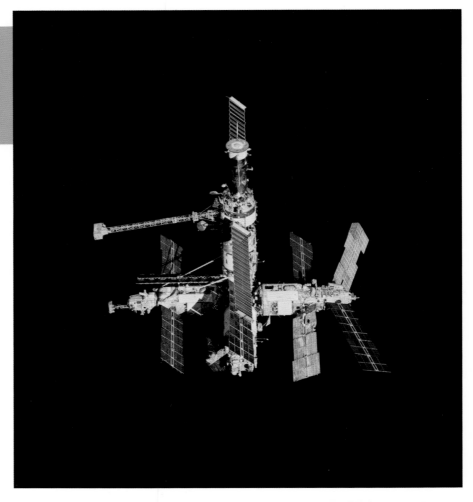

During the STS-120 mission to deliver the *Harmony* node to the International Space Station in 2007, *Discovery*'s crew gained an unexpected task. Astronaut Scott E. Parazynski delicately repaired a damaged solar array. *NASA*

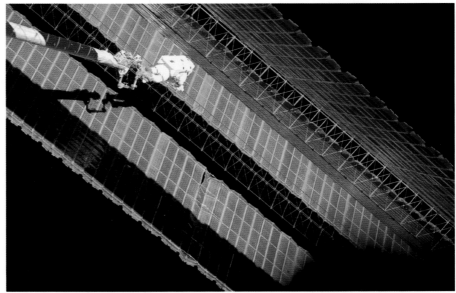

orbiter carried observatories, satellites, or laboratories for scientific research. It also delivered the Sun-circling explorer Ulysses and the Upper Atmosphere Research Satellite. NASA's science missions had several purposes: to exploit microgravity as a laboratory environment, to understand better the changes in humans and other organisms during long stays in space, to take advantage of the clear viewing of Earth and cosmos from above the atmosphere, and to pursue both basic and applied science in quest of benefits for people on Earth.

Discovery's science missions covered a range of disciplines—primarily Earth and atmospheric observations, materials processing, biology, and biomedicine. Several times it carried a Spacelab or SPACEHAB laboratory module in the payload bay, where scientists worked in shifts around the clock. Other times it carried a platform loaded with largely automatic devices, or it released and recovered a small free-flying satellite for particular experiments. *Discovery* made two "Mission to Planet Earth" flights among its science missions in the 1990s.

Discovery opened 1995 with the first of nine shuttle missions to the Russian space station Mir in a cooperative program to prepare for International Space Station partnership. *Discovery* made the first Mir close approach and fly-around, and in 1998 it completed the final Mir docking mission, returning the last of seven American astronauts who alternately lived on the space station. For the Shuttle-Mir missions the Russian and U.S. space agencies shared training, crews, orbital operations, and mission control responsibilities to set the stage for their roles as principal space station partners in the new century. This series of missions included a number of firsts, several of them on *Discovery*.

From 1999 on, all but one of *Discovery*'s last fourteen missions went to the International Space Station. *Discovery* made the first docking with the nascent station to prepare it for its first resident crew and paid its final visit to the station, filled with supplies. Its primary roles in space station assembly were deliveries of

STS-114 Commander Eileen M. Collins put *Discovery* through the rendezvous pitch maneuver, or backflip, for the first time upon approaching the International Space Station in 2005. Cameras and crew on the station surveyed the tiles for signs of damage. *NASA*

The crews of *Discovery*'s STS-120 mission (in red) and International Space Station Expedition 16 (in blue) pause from work to demonstrate weightlessness inside the newly installed *Harmony* node. This crew portrait, like most during the shuttle era, reflects the new demographics of spaceflight. It also captures the first time two women held simultaneous command of space missions: Peggy A. Whitson (lower row) and Pamela A. Melroy (middle row, center). *NASA*

CLOCKWISE FROM TOP, LEFT: The public delighted in *Discovery*'s arrival cruise over Washington, D.C., and surrounding areas on April 17, 2012. The shuttle's history was shaped by political debates and decisions in the nation's capital. *NASA*

The Museum's "Welcome, *Discovery*" ceremony on April 19, 2012, brought together two venerable orbiters, a military band, astronauts, dignitaries, and space enthusiasts to mark the historic occasion. *NASM*

Discovery is on permanent display in the McDonnell Space Hangar at the Museum's Stephen F. Udvar-Hazy Center in Chantilly, Virginia, about twenty-five miles from Washington, D.C. It is seen by more than a million visitors each year. *NASM*

long truss segments, the *Harmony* connecting node, Japan's *Kibo* laboratory module, and the Leonardo Multipurpose Logistics Module. Most of its missions also included space station crew exchanges. *Discovery* flew one more mission to the space station than the other orbiters; its only other mission in this period went to service the Hubble Space Telescope.

Discovery was selected for two return-to-flight missions after the 2003 *Columbia* accident. Both demonstrated a variety of new procedures for improved ability to detect and repair damage in orbit, and thus reduce the risk of another mishap. On the first of the two flights, commander Eileen M. Collins put the orbiter through a rendezvous pitch maneuver, or backflip, in view of the space station crew and cameras so they could look for tile or wing damage. Her crew also accomplished the first EVA underneath the orbiter, using an extended manipulator arm to position an astronaut for close inspection of tiles. The second mission, commanded by Steven W. Lindsey, who also commanded *Discovery*'s final mission, repeated these inspection tasks and demonstrated inflight repair techniques for damaged tiles and wings. After these successes, shuttle flights resumed assembling the International Space Station.

In 2004, President George W. Bush announced that the Space Shuttle program would end upon completion of the station, and NASA soon began planning to retire the orbiters. Each vehicle made a final flight in 2011 to put finishing touches on the

station, but the missions had the air of victory laps. Crew size was smaller, many of the crew members were logging their first flights, and the payloads were primarily stockpiles of supplies to meet the station's needs for a foreseeable future without a large cargo vehicle to resupply it. On its final trip in orbit *Discovery* delivered the astronauts' assistant, Robonaut 2, to the station.

Discovery's crews for its thirty-nine missions reflected the diversity of the astronaut corps in the Space Shuttle era. Its thirty-two commanders (six of whom did so more than once) included the first female pilot and both female commanders, the first African American commander, and another African-American commander who later became NASA administrator. The 184 astronauts, cosmonauts, and payload specialists who flew on *Discovery* included twenty-eight women and citizens of Canada, France, Germany, Italy, Japan, Russia, Saudi Arabia, Spain, Sweden, and Switzerland. The first Hispanic woman, the first Canadian woman, the first Asian American, the first African-American spacewalker, the first astronauts from Spain and Sweden, the first senator, the first commercial and military payload specialists, and the only Mercury astronaut to fly on the shuttle all made their firsts on this orbiter.

Discovery's final voyage began in Florida and ended in the nation's capital on April 17, 2012. NASA assigned OV-103 to the Smithsonian National Air and Space Museum, and, when the orbiter was properly configured for public display, mounted it on the Boeing 747 Shuttle Carrier Aircraft for the journey to its permanent home. After a spectacular fly-over of the Washington, D.C., metropolitan area that brought crowds pouring out of homes, schools, and office buildings to witness the sight, the paired craft landed at Dulles International Airport in suburban Virginia.

Two days later, *Discovery* arrived at the Museum's Stephen F. Udvar-Hazy Center for a ceremonial transfer. Escorted by more than thirty of its mission commanders and crewmembers and greeted by Sen. John Glenn, who had returned to space on *Discovery* in 1998, the longest serving orbiter stood nose-to-nose with its prototype, OV-101 *Enterprise*. This historic meeting of the two vehicles stirred both celebration and nostalgia for the decades of human spaceflight in a reusable spaceplane. Nothing like the Space Shuttle had flown before, and it seemed likely that nothing similar would fly again. The final flight of *Discovery* into retirement signaled the end of the Space Shuttle era.

Now *Discovery* occupies the center of the Museum's space hangar, permanently landed amid artifacts that shaped the history of spaceflight from Robert Goddard's experimental rockets to the engines that propelled launch vehicles higher and the capsules that carried men farther. Nearby, other shuttle-era artifacts relate to *Discovery*'s history: a Canadarm, a Manned Maneuvering Unit, a TDRS satellite model, Spacelab, and smaller items of crew equipment. In retirement, *Discovery* remains an icon of American achievements in space and the hope of making spaceflight routine. It is also a reminder that spaceflight remains a costly and risky venture. It is not yet clear what future spacecraft and missions will emerge, but *Discovery* will long stand as eloquent witness for the Space Shuttle era.

—*Valerie Neal*

DISCOVERY
SPECIFICATIONS

MANUFACTURER:
Rockwell International,
prime contractor
FORWARD FUSELAGE AND CREW CABIN: Rockwell International
MID FUSELAGE:
General Dynamics
PAYLOAD BAY DOORS:
Rockwell International
WINGS: Grumman
AFT FUSELAGE:
Rockwell International
VERTICAL STABILIZER:
Fairchild Republic
MAIN ENGINES:
Rockwell International,
Rocketdyne Division
ORBITAL MANEUVERING SYSTEM:
McDonnell Douglas
LENGTH: 122 ft. (37m)
WINGSPAN: 78 ft. (24m)
HEIGHT: 57 ft. (17m)
ORBITAL ALTITUDE RANGE: 115 to 400 statute mi. (185 to 644 km)
MAXIMUM CARGO TO ORBIT:
63,500 lbs. (28,803 kg)
HEAVIEST WEIGHT AT LAUNCH:
STS-124: 269,123 lbs.
(122,072 kg)
WEIGHT (CURRENT): 161,325 lbs.
(73,176 kg)

DISCOVERY MISSION LOG

Flight	Mission	Type	Crew	Historic Distinctions and Firsts
1	STS-41D 1984 Aug. 30–Sept. 5	Comm. Satellites: SBS, SYNCOM, TELSTAR	Hartsfield, Coats, Resnik, Hawley, Mullane, C. Walker	First mission to deploy three satellites; First commercial payload specialist in space, Charles D. Walker of McDonnell Douglas
2	STS-51A 1984 Nov. 8–16	Comm. Satellites: TELESAT, SYNCOM	Hauck, D. Walker, A. Fisher, Gardner, J. Allen	First mission to retrieve and return inoperable satellites (WESTAR and PALAPA); Third and final use of Manned Maneuvering Unit
3	STS-51C 1985 Jan. 24–27	Dept. of Defense: Classified	Mattingly, Shriver, Onizuka, Buchli, Payton	First dedicated national security mission and first DOD crew member, Gary E. Payton; Ellison S. Onizuka, first American of Asian- Pacific ancestry in space
4	STS-51D 1985 April 12–19	Comm. Satellites: TELESAT, SYNCOM	Bobko, D. E. Williams, Seddon, Hoffman, Griggs, C. Walker, Garn	Senator E. Jake Garn, first U.S. government official in space
5	STS-51G 1985 June 17–24	Comm. Satellites: ARABSAT, TELSTAR, MORELOS	Brandenstein, Creighton, Lucid, Fabian, Nagel, Baudry, Al-Saud	Sultan Salman Abdulazziz Al- Saud, first Saudi, first Arab, and first member of royalty in space
6	STS-51I 1985 Aug. 27–Sept. 3	Comm. Satellites: AUSSAT, ASC, SYNCOM	Engle, Covey, van Hoften, Lounge, W. Fisher	Crew retrieved, repaired, and redeployed the SYNCOM satellite deployed on STS-51D
7	STS-26 1988 Sept. 29–Oct. 3	NASA Satellite: Tracking and Data Relay Satellite (TDRS)	Hauck, Covey, Lounge, Hilmers, G. Nelson	Return to Flight mission after *Challenger* accident
8	STS-29 1989 March 13–18	NASA Satellite: TDRS	Coats, Blaha, Bagian, Buchli, Springer	
9	STS-33 1989 Nov. 22–27	Dept. of Defense: Classified	F. Gregory, Blaha, Musgrave, Carter, K. Thornton	Frederick D. Gregory, first African American to command a space mission
10	STS-31 1990 April 24–29	Science Satellite: Hubble Space Telescope	Shriver, Bolden, Hawley, McCandless, Sullivan	Higher orbit than any previous shuttle mission: 380 miles (612 km)
11	STS-41 1990 Oct. 6–10	Science Satellite: Solar Probe Ulysses	R. Richards, Cabana, Shepherd, Melnick, Akers	

Flight	Mission	Type	Crew	Historic Distinctions and Firsts
12	STS-39 1991 April 28–May 6	Dept. of Defense: Partly Classified	Coats, Hammond, Bluford, Harbaugh, Hieb, McMonagle, Veach	
13	STS-48 1991 Sept. 12–18	Science Satellite: Upper Atmosphere Research Satellite	Creighton, Reightler, Buchli, Gemar, M. Brown	A "Mission to Planet Earth"
14	STS-42 1992 Jan. 22–30	Science: Spacelab International Microgravity Lab-1	Grabe, Oswald, Thagard, Hilmers, Readdy, Bondar, Merbold	Roberta L. Bondar, first Canadian woman in space
15	STS-53 1992 Dec. 2–9	Dept. of Defense: Partly Classified	Walker, Cabana, Bluford, J. S. Voss, Clifford	Last national security shuttle mission
16	STS-56 1993 April 8–17	Science: Spacelab Atmospheric Laboratory (ATLAS-2)	Cameron, Oswald, Foale, Cockrell, Ochoa	Ellen Ochoa, first Hispanic woman in space; A "Mission to Planet Earth"
17	STS-51 1993 Sept. 12–22	Comm. Satellite: ACT-TOS Science Satellite: ORFEUS-SPAS	Culbertson, Readdy, Newman, Bursch, Walz	
18	STS-60 1994 Feb. 3–11	Science: SPACEHAB-2, Wake Shield Facility	Bolden, Reightler, Chang-Diaz, Davis, Sega, Krikalev	Sergei K. Krikalev, first Russian cosmonaut to serve on crew of a U.S. space mission; Charles F. Bolden Jr., second African-American spacecraft commander
19	STS-64 1994 Sept. 9–20	Science: Lidar Technology Experiment (LITE)	R. Richards, Hammond, Linenger, Helms, Meade, Lee	First untethered spacewalk since 1984 as astronauts tested SAFER emergency backpack
20	STS-63 1995 Feb. 3–11	Mir and Science: SPACEHAB-3	Wetherbee, Collins, Harris, Foale, J. E. Voss, Titov	Eileen M. Collins, first female shuttle pilot; First of nine Shuttle-Mir missions (rendezvous without docking); Bernard A. Harris, Jr., first African American spacewalker; *Discovery*, first orbiter to complete 20 missions
21	STS-70 1995 July 13–22	NASA Satellite: TDRS	Henricks, Kregel, D. Thomas, Currie, Weber	Last TDRS deployment

DISCOVERY MISSION LOG

CONTINUED

Flight	Mission	Type	Crew	Historic Distinctions and Firsts
22	STS-82 1997 Feb. 11–21	Hubble Space Telescope Servicing	Bowersox, Horowitz, Tanner, Hawley, Harbaugh, Lee, S. Smith	Second Hubble servicing mission; crew replaced two large science instruments and ten degraded components during five EVAs (spacewalks); Highest shuttle flight then to date: 385 miles (620 km)
23	STS-85 1997 Aug. 7–19	Science Satellite: CRISTA-SPAS	C. Brown, Rominger, Davis, Curbeam, Robinson, Tryggvason	
24	STS-91 1998 June 2–12	Mir and Science: SPACEHAB-3	Precourt, Gorie, Chang-Diaz, Lawrence, Kavandi, Ryumin, A. Thomas	Final Shuttle-Mir docking mission; Andrew S. W. Thomas, last U.S. astronaut returned from stay on Mir
25	STS-95 1998 Oct. 29–Nov. 7	Science: SPACEHAB and SPARTAN	C. Brown, Lindsey, Robinson, Parazynski, Duque, Mukai, Glenn	John Glenn's return to space at age 77, the only original Mercury astronaut to fly on the shuttle; *Discovery*, first orbiter to fly 25 times
26	STS-96 1999 May 27–June 6	International Space Station (ISS) Assembly	Rominger, Husband, Jernigan, Ochoa, Barry, Payette, Tokarev	First docking with nascent space station to prepare it for first resident crew; installed two cranes; 45th spacewalk in shuttle history
27	STS-103 1999 Dec. 19–27	Hubble Space Telescope Servicing	C. Brown, S. Kelly, S. Smith, Clervoy, Grunsfeld, Foale, Nicollier	Third Hubble servicing mission; crew upgraded telescope for its second decade of operations
28	STS-92 2000 Oct. 11–24	ISS Assembly: Z1 Truss and Pressurized Mating Adapter	Duffy, Melroy, Chiao, W. McArthur, Wisoff, Lopez-Alegria, Wakata	100th Space Shuttle mission
29	STS-102 2001 March 8–21	ISS Logistics: *Leonardo* Module and Crew Exchange	Wetherbee, J. Kelly, A. Thomas, P. Richards; Up: J. S. Voss, Helms, Usachev; Down: Shepherd, Gidzenko, Krikalev	First ISS crew exchange; Longest spacewalk in shuttle history, 8 hours 56 minutes by Susan J. Helms and James S. Voss
30	STS-105 2001 Aug. 10–22	ISS Logistics and Crew Exchange: *Leonard* Module	Horowitz, Sturckow, Forrester, Barry; Up: Culbertson, Dezhurov, Tyurin; Down: J. S. Voss, Helms, Usachev	*Discovery*, first orbiter to fly 30 times

For mission details, see www.spaceflight.nasa.gov or http://www.nasa.gov/mission_pages/shuttlemissions

Flight	Mission	Type	Crew	Historic Distinctions and Firsts
31	STS-114 2005 July 26–Aug. 9	Return to Flight and ISS Logistics	Collins, J. Kelly, Noguchi, Robinson, A. Thomas, Lawrence, Camarda	First rendezvous pitch maneuver (orbiter backflip) for inspection by ISS crew, prompted by *Columbia* accident ; First EVA underneath the orbiter
32	STS-121 2006 July 4–17	Return to Flight and ISS Logistics Leonardo Module and Crew Addition	Lindsey, M. Kelly, Fossum, Nowak, Wilson, Sellers; Up: Reiter	Further testing of orbiter inspection and in-flight repair techniques prompted by *Columbia* accident; Most photographed shuttle mission then to date
33	STS-116 2006 Dec. 9–22	ISS Assembly and Crew Exchange: P5 Truss	Polansky, Oefelein, Patrick, Curbeam, Fuglesang, Higginbotham; Up: S. Williams; Down: Reiter	Crew rewired the ISS power system
34	STS-120 2007 Oct. 23–Nov. 7	ISS Assembly and Crew Exchange: *Harmony* Module	Melroy, Zamka, Parazynski, Wilson, Wheelock, Nespoli; Up: Tani; Down: Anderson	*Discovery* became only orbiter flown by both women commanders, Eileen M. Collins and Pamela A. Melroy; both also flew as pilots on *Discovery*
35	STS-124 2008 May 31– June 14	ISS Assembly and Crew Exchange: *Kibo* Module	M. Kelly, Ham, Nyberg, Garan, Fossum, Hoshide; Up: Chamitoff; Down: Reisman	Delivered and installed Japan's large laboratory module, *Discovery*'s heaviest payload; *Discovery* became only orbiter to fly 35-plus times
36	STS-119 2009 March 15–28	ISS Assembly and Crew Exchange: S6 Truss	Archambault, Antonelli, Acaba, Swanson, Arnold, Phillips; Up: Wakata; Down: Magnus	Delivered and installed final set of ISS solar arrays
37	STS-128 2009 Aug. 28–Sept. 11	ISS Logistics and Crew Exchange: *Leonardo* Module	Sturckow, Ford, Forrester, Hernandez, Olivas, Fuglesang; Up: Stott; Down: Kopra	30th shuttle mission to ISS
38	STS-131 2010 April 5–20	ISS Logistics: *Leonardo* Module	Poindexter, Dutton, Mastracchio, Metcalf-Lindenberger, Wilson, Yamazaki, Anderson	
39	STS-133 2011 Feb. 24–March 9	ISS Assembly: Permanent Multipurpose Module	Lindsey, Boe, Drew, Bowen, Barratt, Stott	Delivered Robonaut 2 astronaut assistant to ISS; 35th shuttle mission to ISS

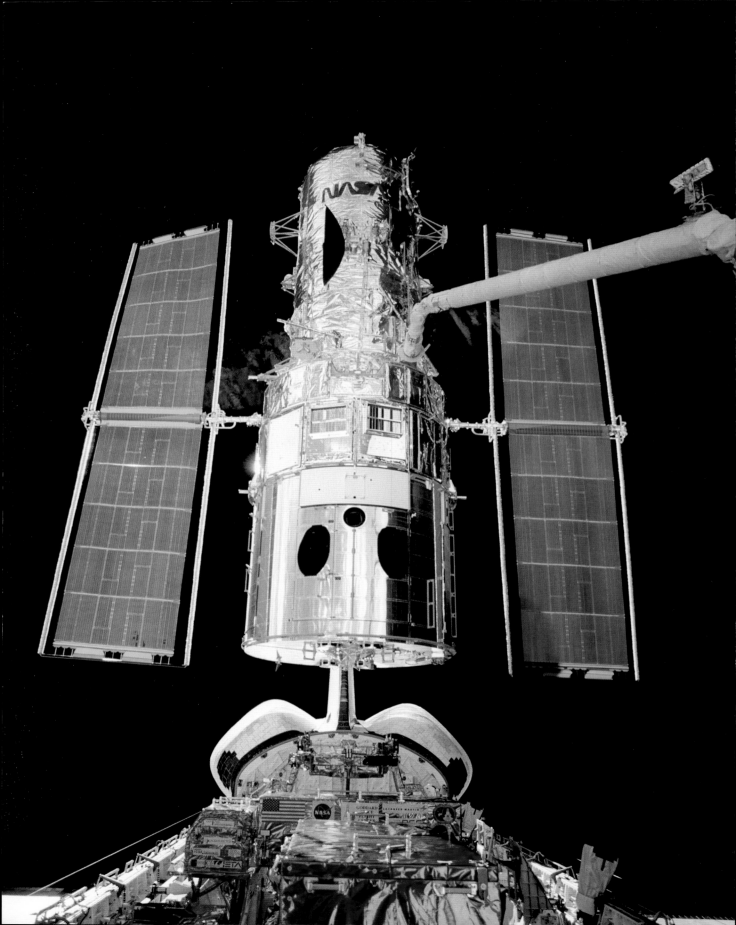

THE HUBBLE SPACE TELESCOPE (HST), at this writing, is still very much alive, and it is far more capable of probing the universe than when it was launched in 1990. Most of its planners, promoters, and builders are still with us, and it continues to produce pathbreaking science because its managers and users are well supported by NASA's astrophysics program. It is not yet purely history, but it is already history.

The HST is the best known astronomical telescope at present. Its popularity comes from the transcendent vision it has provided humanity of the things that lurk in deep space and time, presented in aesthetically compelling ways by teams of devoted astronomers and digital artists. Its productivity comes from the thousands of scientific works based upon its observations that have refined and, in some cases, transformed our understanding of the universe. It is the most expensive astronomical telescope in history, and those who spent decades campaigning for it, designing it, and using it, have been noticeably changed by the experience. Its trials, tribulations, and triumphs have been noticed by the mass media. More than any other telescope in history, it has a huge popular following. It is a living astronomical milestone.

The actual telescope is presently in orbit around the Earth, at a mean distance of some 350 miles, making a circuit (with respect to the stars) every ninety-seven minutes, travelling at some five miles per second. It was carried into space by the Space Shuttle *Discovery* in 1990, and, although *Discovery* has been retired and is

THE HUBBLE SPACE
TELESCOPE
// 11

presently on view at the Smithsonian National Air and Space Museum's Udvar-Hazy Center, there are no plans to bring Hubble back. But we do have important parts of the Hubble, returned from space, or left over from its original construction. To date, the Museum has collected more than fifty objects relating to its building and operation.

The deployment of the Hubble Space Telescope into orbit, on April 25, 1990, from Space Shuttle *Discovery. NASA*

Dreams and Promises

Astronomers and others had long hoped to see the universe unhindered by the ocean of air we live in. Necessary as it is for life, it blurs and blocks the heavens from view. None of this was a problem until the telescope was invented, followed by Galileo turning it to the heavens in 1609, showing that the moving bodies in the sky were physical places with Earthlike features. Only by the late nineteenth century was the blurring of the Earth's atmosphere considered severe enough to compel astronomers to put their telescopes on high mountains, at first to clear much of the atmosphere and then to avoid the lights and smoke spread by modern civilization.

Since then, astronomers have been building bigger and bigger telescopes at high, dry, remote locations.

But some dreamed of removing the atmosphere entirely. Right after World War II, with the German V-2 missile showing what was possible, astronomer Lyman Spitzer described what astronomical problems might be addressed with a large telescope that could be launched by a rocket into orbit. Most prescient, he felt that it would "uncover new phenomena not yet imagined and [would] perhaps modify our basic concepts of space and time." And when it was finally launched in 1990, Spitzer recalled the milestones that brought it to life. His dream remained a "long-range goal" in the 1950s but "got into high gear" after *Sputnik* challenged the world and flung America into space.

For Spitzer, the key technology to worry about was not the launch vehicle. It was not even the telescope; astronomers knew how to build them. It was the means of getting the data the telescope gathered, from its origin in space to the ground. He chaired a 1965 summer study that reported what astronomers hoped to learn with a Large Space Telescope, confident that such a means of detection and transmission of data was at hand. Not everyone was as confident, and so various competing designs persisted for the Large Space Telescope. Some were "man-tended," using film that would be returned by astronauts during resupply missions.

By 1970, however, Spitzer and other astronomers both within and outside NASA were fairly certain that technology developed within the classified intelligence community supported electronic image detection and transmission. Therefore, astronomers, and soon NASA, championed a robotic observatory with an electronically equipped telescope. At first they hoped it would have a three-meter mirror, but by 1974 it was downsized to a 2.4-meter (94-inch) mirror to reduce escalating costs and allow it to fit inside the cargo bay of the planned Space Shuttle. This was an astute political shift to accommodate NASA's centralization around Shuttle operations. By 1975, it was no longer the Large Space Telescope (or Lyman Spitzer Telescope, as some felt it should be) but merely the Space Telescope. It was finally named the Hubble Space Telescope in 1983, to indicate its scientific legacy. Among its many claimed goals, it would extend astronomer Edwin Hubble's 1930s recognition's that galaxies existed, and were all moving away from one another.

Hubble's launch was scheduled for the mid-1980s, but the *Challenger* disaster delayed everything for years. Finally launched in April 1990, astronomers eagerly awaited first word of the wonders it would reveal.

Post-Launch Reality

After a month of launch, deployment, and engineering checkout, astronomers were cautiously optimistic while viewing the first images. The telescope did not yet produce pinpoint images of stars, but they were a significant improvement over what could be done from Earth and, the astronomers hoped, were "a hint of things to come when further alignment and focusing is expected to produce stellar images." But after more weeks of tuning, the images remained soft disks of light surrounded by halos—hardly what astronomers expected or needed from a space telescope. What the images

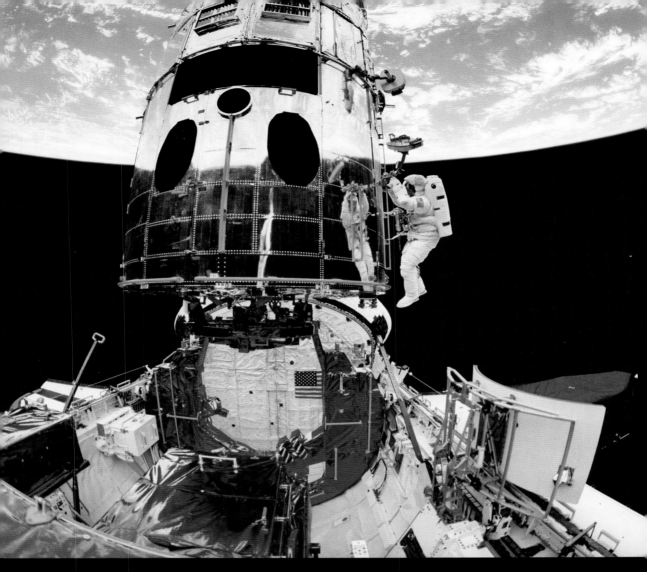

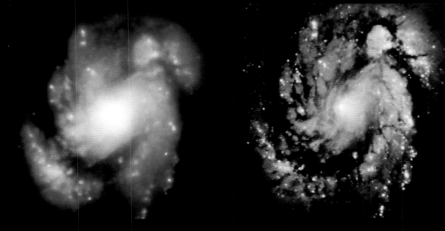

ABOVE: Replacing WFPC during the first servicing mission in 1993. With the original WFPC removed, WFPC2 sitting at the lower right, awaits insertion into the large rectangular aperture in HST, just behind the main mirror. *NASA*

Images of the spiral galaxy M100, taken by the original WFPC and then by the modified WFPC2, showing enhanced resolution after the first servicing mission. *NASA*

told them by June was that the Hubble was incapable of concentrating all the light it received with 100 percent efficiency because its main mirror was imperfectly figured. It was far less efficient, which meant that even though the various non-imaging instruments—the photometers and spectrographs—could still operate and produce science, they would be doing so at a significantly reduced pace. Hubble had to be fixed.

There had been options for how to build and operate what became the Space Telescope. One way was to design, build, test, and fly multiple copies of a satellite. If the first one failed, learn from it and move to the second one. There were some astronomers who wanted to use that model, flying multiple generations of space telescopes, each one better than the last, or at least optimized for a specific purpose. Another way was to build a telescope that could be serviced in orbit somehow, either through on-orbit visits by astronauts or by returning the telescope to Earth. Any mission involving humans, of course, would be far more expensive, unless human spaceflight became economical.

The Hubble Space Telescope embodied both choices. It was designed to be serviced by astronauts, but it was also a product of lessons learned from previous missions, ranging from balloon-borne telescopes to the Orbiting Astronomical Observatories to the International Ultraviolet Explorer.

Other missions of Hubble's generation of "Great Observatories," as NASA called them, also had precursors, optimized for the invisible portions of the vast spectrum of energy that celestial bodies radiate. But none of these missions was serviceable. And none was as complex and expensive as the Space Telescope promised to be. Some astronomers felt that a servicing scenario was critical to insure that the enormous investment would pay off. But before Hubble, only the Solar Maximum Mission was developed with a modular design and a special grapple fixture. It was serviced once by astronauts during a repair mission in April 1984. Thus, the Hubble was the only astronomical satellite capable of being improved during the Shuttle era. Aside from its scientific legacy, this unique operational legacy has been the Smithsonian's most important guide, and enabler, for preserving its material legacy.

Early thinking had been that if it needed repair or periodic upgrade, the Hubble could be returned to Earth, but the emergent plan was for on-orbit servicing. The obvious extra costs and dangers of multiple Shuttle launches in returning and then reflying HST, and the fact that this would not demonstrate a new capability to actually perform complex tasks in space, left only the on-orbit scenario.

Between the revelation in June 1990 of the flawed primary mirror, and the repair (often called "rescue") mission in December 1993, NASA knew its reputation and future were on the line. Staff of the Space Telescope Science Institute in Baltimore, the organization created to manage scientific operations, joined with NASA and telescope contractors to decide to remove one of the instruments and replace it with a device called the Corrective Optics Space Telescope Axial Replacement (COSTAR), a collection of tiny mirrors on deployable stalks that would reverse the errors in the mirror's light beam feeding the scientific instruments. It would work for all the instruments except the Wide Field and Planetary Camera (WFPC), which was the primary imaging device.

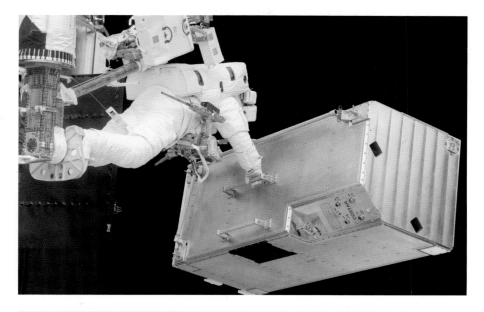

COSTAR was removed from Hubble in 2009 during the final servicing mission. All the other instruments had by then been replaced with internal modifications to correct for the flawed primary mirror. *NASM*

COSTAR contained a set of tiny mirrors on deployable stalks that were inserted into the primary light beam of the HST mirror to send corrected light to all the instruments except the WFPC2. *NASM*

Fortunately, given the rapid advance of detector technology, a second-generation WFPC2 was already under development, and could readily be adjusted to reverse the primary mirror's flaw. In addition to installing those devices, the astronauts on the first servicing mission mounted new solar power arrays that resisted thermal effects caused by the satellite's periodic passage from darkness to sunlight, and they replaced gyroscopes, fuses and electronics. The mission resuscitated the reputation of the space agency and reenergized astronomers worldwide. This mission also gave curators at our Museum hope that they would collect those pieces returned from space.

Servicing the Hubble on five occasions from 1993 to 2009 provided NASA with a powerful argument for human spaceflight's ability to provide critical services. Looking back at the first visit in 1993, NASA's Space Telescope Science Institute rejoiced on its website that: "This successful mission not only improved Hubble's vision—which led to a string of remarkable discoveries in a very short time—but it also validated the effectiveness of on-orbit servicing."

Indeed, all the servicing missions were successful, garnering wide media coverage. The Hubble was not only repaired in 1993, but improved. And in subsequent missions, gyroscopes, computers, and batteries were restored, corrected, or improved. Scientific instruments were replaced, either with upgrades or new instruments with capabilities not available when Hubble was launched.

Preserving the History of the HST

In setting priorities for what to preserve from the Hubble Space Telescope mission, the National Air and Space Museum must consider not only the hardware that gave us the images and insights about the universe but also the devices created to make

A SUPERMASSIVE BLACK HOLE

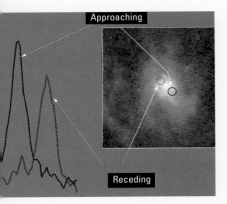

The Faint Object Spectrograph was pointed to different parts of the swirling disk of gas at the center of M87. It revealed that one side is moving in our direction while the other side is equally receding, indicating that the disk is in rapid rotation. *NASA/STSci*

The Hubble's ability to isolate features of very small dimension within large extended objects made it possible to pinpoint and record the spectrum of a tiny central region of the massive galaxy M87, known to be a powerful radio source. M87, the brightest galaxy in a cluster of galaxies spread across the constellation Virgo, is in a class of highly energetic objects called active galactic nuclei (AGN). The galaxy also is notable because it has a jet-like feature, like a thin finger, jutting out from its center. That jet is caused by accelerated gas moving at velocities near the speed of light.

Features like this have led astronomers to long suspect that something extremely energetic, and therefore hugely massive, must be lurking in the hearts of galaxies of the AGN class. The Hubble was turned to the task soon after the astronauts aboard shuttle mission STS-61 installed COSTAR during the first servicing mission. That upgrade restored the Faint Object Spectrograph (FOS) to full power and spatial resolution.

By May 1994, the FOS team of astronomers from the Space Telescope Science Institute, Johns Hopkins University, Applied Research Corporation, and the University of Washington, had collected observations of the spectrum at five locations at the galaxy's core, and two locations outside the core. What they inferred from their spectroscopic measurements was that there is a rapidly rotating disk of high-temperature gases circling the nucleus of the galaxy. Given the speeds of the gases swirling around the center at opposite points of the disk, which varied by more than 1,000 kilometers per second, and the linear size of the disk determined by knowing the distance to the galaxy, there must be an object some three million times the mass of the Sun lurking at the very center. Given so much mass concentrated at the very center of such a small volume of space, the only thing it can be, according to known physics, is what astronomers call a supermassive black hole. These objects can contain millions to billions of solar masses in a volume smaller than the Earth's orbit. Their existence is inferred by the presence of extreme gravitational fields.

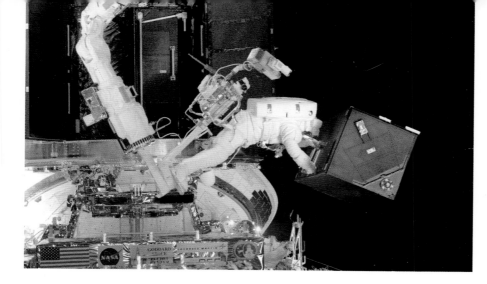

An astronaut anchored to, and maneuvered by, the articulating Canadarm removes the Faint Object Spectrograph from Hubble during the third servicing mission in February 1997. *NASA*

it the only free-flying instrument to be repeatedly serviced and upgraded by humans. Given this complex legacy, what options do we have, and what choices do we make within those options, for preserving this history? Specifically, what among the remains of its material heritage is, or will be, most revealing of its history?

In the early 1980s, historians at The Johns Hopkins University and at the Museum joined forces to create the Space Telescope History Project, a multiyear effort to create a documentary archive of oral history interviews and copies of documentary evidence, including letters, notes, minutes of meetings, internal technical correspondence, memoranda of understanding, and visual evidence in the form of images. Based upon this body of data, Robert W. Smith, who led the effort, prepared the award-winning book *The Space Telescope: A Study of NASA, Science, Technology and Politics* (1989), with the assistance of three other historians.

Throughout this process, Smith, and his colleague Joseph Tatarewicz, were sensitive to the material legacy resulting from building the Space Telescope, namely a full-sized backup ninety-four-inch mirror with the same specifications of the primary mirror that eventually flew on the Hubble, and a full-scale mock-up of the physical satellite, called the Structural Dynamic Test Vehicle (SDTV). Meanwhile, NASM's curator of the history of space astronomy, David DeVorkin, devoted his attention to procuring high-definition models and elements of the instruments and detectors that were to be flown. His primary attention—in the early 1980s, at least—was to build a new astronomy gallery, titled *Stars: From Stonehenge to the Space Telescope*. As part of this effort, he collected 1/5th scale models of the optical telescope assembly from the prime contractor, the Perkin-Elmer Corporation, and a 1/5th scale model of the spacecraft that would house the assembly, from the Lockheed Corporation. *Stars* opened in June 1983, and museum labels anticipated that the HST would soon be launched. The Museum already had smaller models of the Space Telescope, starting with a 1/40th scale model of the original LST design acquired from the Marshall Space Flight Center in 1975.

By then the Museum had acquired the SDTV from Lockheed, and restored it to represent its operational appearance *circa* 1976. It was displayed in Space Hall from 1989 until 1996 when it was removed and modified to look like the HST in orbit and displayed again starting in 1997. Lockheed also donated an assortment of test parts that had been used at one time or another with the SDTV, notably a test example of one of Hubble's critical reaction-wheel assemblies, part of the pointing and guiding system.

As might be expected, the Museum's collection of Hubble hardware grew with the deployment and servicing missions. In 1997, NASM acquired Shuttle astronaut Kathryn D. Sullivan's Velcro-backed leather nametag from her spacesuit on the HST deployment mission (STS-31, *Discovery*, 1990). After the second servicing mission in 1997, the Museum acquired the Faint Object Spectrograph (FOS), one of the original instruments launched with HST that in 1994 provided confirming evidence for a supermassive black hole in the giant elliptical galaxy M87. The FOS was not defective; it was switched out to provide room for a new instrument, the Space Telescope Imaging Spectrograph (STIS). On the same mission, the Goddard High Resolution Spectrograph was replaced by a powerful infrared imaging camera and spectrograph, the Near Infrared Camera

The Hubble Space Telescope as depicted in NASM's Space Hall. The original Structural Dynamic Test Vehicle (SDTV) acquired by the Museum was later covered with thermal blankets, handrails, and a set of fake solar panels to depict the HST deployment from the Space Shuttle. *NASM*

and Multi-Object Spectrometer (NICMOS), which could probe the deepest depths of space. These replacements continued a standard practice in modern astronomy and astrophysics: making the broadest use of a large, expensive telescope by switching out cameras and other devices that are capable of performing different tasks. The second servicing crew also replaced a failed fine-guidance sensor and reaction wheel assembly and installed a new solid-state data recorder that was faster and had larger capacity.

A third mission was hastily organized when it was feared that the fourth of Hubble's six gyroscopes might fail. Hubble needs three gyroscopes to observe a target, so this was a crisis that led to shutting the telescope down and closing the hatch. Without at least three gyros, it was conceivable that Hubble would drift, bringing the Earth, Moon, or—worst case—the Sun into view. That much light would permanently damage the sensitive system through the telescope. In December 1999, the third-mission astronauts replaced all six gyroscopes and a fine guidance sensor, applied patches to the thermal covering, and installed new batteries and a more powerful central computer. "Hubble was as good as new," declared the NASA website.

During the 1990s, the Museum was again planning a new astronomy gallery to replace *Stars*. The new theme was the history of cosmology. Naturally, Hubble would be center stage, both the telescope and the man who had made the two signal

GOING DEEP, ULTRA DEEP, AND EXTREMELY DEEP

The director of the Space Telescope Science Institute in Baltimore is given discretionary observing time on HST as partial compensation for running the large organization responsible for managing mission science. Astronomer Robert Williams, who became the Institute's second director in 1993 (succeeding Riccardo Giacconi), tried something different. He and an advisory committee decided to point the telescope to a lonely dark spot in space not affected by scattered light from bright objects. This meant they could take very long exposures without worrying about fogging the images. The spot they chose was in the constellation of the Big Dipper.

They hoped it was empty enough of nearby gas, dust, stars, and galaxies that it would give them a window on the deepest realms of the universe.

Thus was born the Hubble Deep Field, taken with WFPC2 and issued in 1996. It is a composite of 342 exposures taken between December 18 and 28, 1995, and covers an area of the sky only 1/100th of that occupied by the full Moon. Most of the three thousand objects found in the exposure turned out to be highly distant galaxies. The image revealed the collisional processes through which galaxies grew in the early universe.

The revelations of the Hubble Deep Field prompted astronomers to push even deeper, which meant reaching even earlier epochs. As we look into space, we also look back in time. Light takes eight minutes to reach us from the Sun, so what we see is eight minutes old. For the stars, it is years to millennia; for galaxies, eons. The Deep Field took us back some twelve billion years, less than two billion years after the Big Bang. A counterpart in another part of the sky was completed in 1998 with WFPC2, aided by two other instruments. It confirmed that the universe looks the same in both directions. By early 2004, the newly installed and improved Advanced Camera for Surveys went even deeper in Fornax between September 24, 2003, and January 16, 2004. This was the Ultra-Deep Field, revealing some ten thousand galaxies as deep as thirteen billion light years, at a time when the first stars had existed for only a few hundred million years. In 2012, astronomers compiled an even deeper survey, to 13.2 billion years, using accumulated observations over a decade, including those by the Wide Field Camera 3 installed on the last servicing mission in 2009.

As Hubble had its vision clarified and its detectors improved, it repeatedly broke its own records for detecting the most distant, and youngest, objects in the universe. This is a typical example from 2011, revealing a reddish galaxy catalogued as UDFj-39546284 that is estimated to be some 13.2 billion light years away. *NASA, ESA, G. Illingworth [University of California, Santa Cruz], R. Bouwens [University of California, Santa Cruz, and Leiden University], and the HUDF09 Team*

NASA prudently contracted with two firms to produce the flight mirror and an identical backup in case the flight mirror was damaged during manufacture. The backup mirror, polished and figured by Kodak Precision Optics, was perfect. It is presently on display at the National Air and Space Museum in the *Explore the Universe* gallery. *NASM*

Original, flown CCD detector in the NASM Collection. It is an 800 × 800 pixel array, smaller than those found in typical digital cameras today. The postage stamp-sized detector provided small area coverage and required that the WFPC camera use arrays of these detectors. Because their support electronics were large, they had to be separated in the camera. *NASM*

discoveries in the early twentieth century about how the universe is structured. Smith provided leadership and guidance, setting the stage to acquire the backup mirror to the Hubble, which was produced by Kodak and in storage at Perkin-Elmer. DeVorkin concentrated on the instruments, arranging for elements from the first WFPC to be reinstalled within an engineering test unit from the Goddard Space Flight Center. His goal was to use it to illustrate how the original WFPC (as well as WFPC2) overcame the intrinsic size limitations of the charge-coupled devices (CCDs). FOS, which had been on display in Space Hall, was moved into the new gallery as an example of the type of instruments that have been employed in the search for dark matter.

Explore the Universe opened in September 2001, but within a few years, the extraordinary impact of the Hubble images from WFPC2 created by the Hubble Heritage Team at the Institute called for more Museum exposure. Curators covered the western wall of Space Hall with Hubble's greatest images, which included new images from a new Advanced Camera for Surveys (ACS), with CCD chips that were larger than ever, installed during the fourth servicing mission in March 2002. That mission also gave the NICMOS a new cooling system, installed more efficient solar panels, and replaced another faulty reaction wheel. The spectacular images kept flowing, from WFPC2 and ACS.

By late 2003, continuously aging components worried Hubble program leaders as they planned for a fifth servicing mission in 2005. But in January 2004, after the loss of the shuttle *Columbia* in 2003, NASA's leadership canceled the fifth manned mission, tentatively opting for a robotic mission. Astronomers howled, and even the public was shocked—testimony to the popularity of Hubble's images. Under pressure, a new

HUBBLE'S KEY PROJECT

In the 1920s, Edwin Hubble not only discovered that galaxies exist beyond our Milky Way but also that we are in a universe of galaxies. Moreover, they tend to move away from one another according to the relation that, the more distant they are from each other, the faster they are moving. In other words, the universe is expanding—not static. There must have been a time when it started this expansion, an instant when space and time began. Astronomers have come to call that moment the Big Bang.

Hubble was not the only astronomer thinking about the implications of expansion, but he was the first to execute a systematic program of measuring galaxy velocities and independently determining their distances, which led to their correlation. Ever since Hubble's time, the challenge has been to refine the universe's rate of expansion, what astronomers call the Hubble Constant. Has it indeed been constant over time, and just what is its numerical value? If we know just the numerical value, we can estimate the age of the universe. That was the intent of the Hubble Key Project.

As the figure illustrates, astronomers' estimates of the Hubble Constant have changed over time. The vertical lines indicate the range of uncertainty in the numerical value, which indicates uncertainties in the expansion age of the universe. Before the 1950s, the expansion rate was very high, so the expansion age appeared to be far less than the geological age of the Earth, or the ages of stars, based upon theories of nuclear fusion. Hubble's initial value for the expansion rate was some 550 kilometers per second for every million parsecs (a parsec is 3.26 light years), but subsequent recalibrations reduced its numerical value to the point where relative ages fit quite well.

Through the 1960s and 1970s, however, there was still a question over what the value should be, though by then it had been reduced to under 150 kilometers per second for every million parsecs, but the range of uncertainty was not better than 30 percent. That is why one of the key projects assigned to the Hubble Space Telescope was to further refine the correlation of velocity and distance based on variable stars known as Cepheids, as well as other indicators of distance on extragalactic scales. A team of thirteen astronomers proposed the Hubble Key Project in the 1980s. The goal was to beat down the errors to less than 10 percent of the Hubble Constant. The high resolution of the instruments aboard Hubble provided the capability, and so by 1999, the Key Project Team met that goal, recalibrating all the distance indicators to refine the extragalactic distance scale to 71 +/- 7, with a corresponding expansion age of the universe of just over 13 billion years.

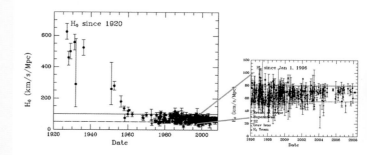

Refinement of the Hubble Constant from Hubble to Hubble

From the 1930s through 2008, astronomers refined knowledge of the Hubble Constant, reducing it from hundreds of km/sec/ megaparsec to a few dozen, thereby increasing the estimated expansion age of the universe. The vertical bars (ranges of error) are also much smaller in the modern estimates. Hubble's "Key" Project played a major role in this refinement. *John P. Huchra, © 2008*

RIGHT: Using WFC3 as the primary imaging camera on Hubble, astronomers peered into the vast clouds of gas and dust in the direction of the Carina Nebula in the southern sky. A star just in the process of forming can be seen in the deep infrared view in the lower frame. *NASA, ESA and the Hubble SM4 ERO Team*

WFC3/UVIS

ABOVE: Astronauts prepare to remove WFPC2 during the last servicing mission, SM4. The white radiator for WFPC2 was exposed to space for some seventeen years, and so became a historical record of the impacts of space debris on the Hubble during that time. For that reason, the surface of the radiator was sent first to the Goddard Space Flight Center and then to the Johnson Space Center, where it was subjected to a thorough examination to ascertain the effects and estimate the degree of microscopic pollution of low Earth orbit by artificial debris. *NASA*

NASA administrator reinstated a final servicing mission, which took place in May 2009. It included the repair of two major instruments (ACS and STIS), the replacement of WFPC2 by a camera with an even larger CCD chip—Wide Field Camera 3, which can record near infrared light as well as visible and near ultraviolet light—and the removal of COSTAR to install the Cosmic Origins Spectrograph. They also installed all new batteries and gyros, a new Fine Guidance Sensor, new thermal blanketing, and a new command and data handling unit. The astronauts attached a grappling fixture in case a future robotic mission might visit, either for repair, returning it to Earth (not likely), or (most likely) for a safe deorbit that would result in the telescope burning up on reentry.

COSTAR, which had been the original correcting device for most instruments, was replaced because all instruments now embodied the optical fix for the mirror problem. The Museum borrowed it from NASA in September 2009. It has now been formally accessioned and is on exhibit in Space Hall.

WFPC2 was briefly exhibited for several months in 2009–10, but it was returned to NASA for evaluation. The radiator unit forming the exposed outer skin of the axial bay had been exposed to space conditions for some sixteen years, and engineers wanted to microscopically examine every square millimeter of its surface to inventory all the space junk that had hit the satellite. Their findings are still pending, but when WFPC2 is ready to be acquired by the Museum, one of its stories will be that of its long exposure to space.

Reflections

The impact of the Hubble Space Telescope on astronomy, not only on astronomers' concepts of the many weird and wonderful ways the universe works but of their sense of what it means to be a practicing astronomer, has been enormous. It easily qualifies as a milestone in the history of astronomy. But its history no doubt will be rewritten more than once.

Looking back over the history of the Hubble Space Telescope, one has to ask if designing it as an instrument that was to be launched by the Space Shuttle, and then serviced by astronauts in subsequent missions, was the right approach. The answer will depend upon what "right" means. The Hubble mission aided science and supported the development of human spaceflight. Therefore, assessing the cost-effectiveness of the mission in terms of the knowledge gained about the astronomical universe would be too narrow. One must assess the cost of the mission in terms of the larger effort, providing the capability of constant servicing in orbit, in this case by the Space Shuttle.

Considering the number of things that went wrong with the telescope, which was obviously seriously flawed at launch, it would not have provided the wealth of information that continues to flow into astronomers' computers without repeated repair. Thus, one might argue that making it serviceable was the right decision. But one must also consider the fact that although none of the other Great Observatories could be serviced, each of them has worked beautifully, returning revolutionary information about the universe. They were placed in orbits that were beyond the capacity of the Shuttle to service, but this also maximized their scientific powers. And their costs, both developmental and operational, were orders of magnitude less, if one factors in the cost of each servicing mission.

The Hubble could have been placed in a higher orbit with less ultraviolet interference from the residual outer atmosphere, but then it could not have been serviced. So the question is, if the Hubble was the only one of the Great Observatories that could be serviced, why was it the only one that needed servicing? Robert Smith observes that by making the telescope serviceable, contractors were able to convince NASA that they could save money by reducing the level of testing for each component. This would, they admitted, increase risk of failure, but they argued that they could fix any failure with on-orbit servicing. And in the 1970s, the predicted cost of on-orbit servicing by the Shuttle was orders of magnitude less than what the actual costs turned out to be.

Smith has indicated that for the extra cost of the servicing missions, a whole new second telescope could have been built. Offhanded speculation by well-informed astronomers have increased this number. Thus, the question is still open whether the Shuttle-centered mission was the right choice for science, though it evidently was for NASA.

—*David H. DeVorkin*

HST FLIGHT UNIT
SPECIFICATIONS

MANUFACTURERS:
Lockheed-Martin Space Systems; Perkin-Elmer Corporation.
LENGTH: 43.5 ft. (13.2m)
DIAMETER (AT BASE): 14 ft. (4.2m)
WEIGHT (AT LAUNCH):
24,500 lbs. (11,110 kg)
PRIMARY MIRROR: 94.5 in. (2.4m); 1,825 lbs. (828 kg)
ORBIT: 307 nautical mi. (569 km, or 353 mi.), inclined 28.5° to the Earth's equator
LAUNCH: April 24, 1990, from Space Shuttle *Discovery* (STS-31)
SERVICING MISSION 1:
December 1993
SERVICING MISSION 2:
February 1997
SERVICING MISSION 3A:
December 1999
SERVICING MISSION 3B:
February 2002
SERVICING MISSION 4:
May 2009

SUGGESTIONS FOR FURTHER READING

General Space History

Burrows, William E. *This New Ocean: The Story of the First Space Age*. New York: Random House, 1998.

Collins, Martin, ed. *After Sputnik: 50 Years of the Space Age*. New York: Collins/Smithsonian Books, 2007.

McDougall, Walter A. *. . . the Heavens and the Earth: A Political History of the Space Age*. New York: Basic Books, 1985.

Neufeld, Michael J. *Von Braun: Dreamer of Space, Engineer of War*. New York: Alfred A. Knopf, 2007.

Siddiqi, Asif A. *Challenge to Apollo: The Soviet Union and the Space Race, 1945–1974*. Washington: NASA, 2000.

Mercury Capsule *Friendship* 7

Glenn, John, with Nick Taylor. *John Glenn: A Memoir*. New York: Bantam, 1999.

Swenson, Loyd S., Jr., James M. Grimwood, and Charles C. Alexander. *This New Ocean: A History of Project Mercury*. Washington: NASA, 1966.

Wolfe, Tom. *The Right Stuff*. New York: Farrar Straus Giroux, 1979.

Telstar

Schwoch, James. *Global TV: New Media and the Cold War, 1946–69*. Champaign: University of Illinois Press, 2008.

Whalen, David J. *Origins of Satellite Communications, 1945–1966*. Washington: Smithsonian Institution Press, 2002.

Corona KH-4B Camera

Clausen, Ingard, and Edward A. Miller. *Intelligence Revolution 1960: Retrieving the Corona Imagery that Helped Win the Cold War*. Chantilly, Va.: National Reconnaissance Office, 2012.

Dwayne A. Day, John M. Logsdon, and Brian Latell, eds., *Eye in the Sky, the Story of the Corona Spy Satellite*. Washington: Smithsonian Institution Press, 1998.

Ruffner, Kevin C., ed., *Corona: America's First Satellite Program*. Washington: Central Intelligence Agency, 1995.

The F-1 Engine

Bilstein, Roger E. *Stages to Saturn: A Technological History of the Apollo/Saturn Launch Vehicles*. Washington: NASA, 1980.

Hunley, J. D. *U.S. Space-Launch Vehicle Technology: Viking to Space Shuttle*. Gainesville: University Press of Florida, 2008.

Kraemer, Robert S. *Rocketdyne: Powering Humans into Space*. Reston, Va.: American Institute of Aeronautics and Astronautics, 2006.

Young, Anthony H. *The Saturn V F-1 Engine: Powering Apollo into History*. New York: Springer, 2009.

Lunar Module LM-2

Brooks, Courtney G., James M. Grimwood, and Loyd S. Swenson, Jr. *Chariots for Apollo: A History of Manned Lunar Spacecraft*. Washington: NASA, 1979.

Kelly, Thomas J. *Moon Lander: How We Developed the Apollo Lunar Module*. Washington: Smithsonian Institution Press, 2001.

Murray, Charles, and Catherine Bly Cox. *Apollo: The Race to the Moon*. New York: Simon & Schuster, 1989.

Neil Armstrong's Spacesuit

Hansen, James R. *First Man: The Life of Neil A. Armstrong*. New York: Simon & Schuster, 2005.

Jenkins, Dennis R. *Dressing for Altitude: U.S. Aviation Pressure Suits—Wiley Post to Space Shuttle*. Washington: NASA, 2012.

Thomas, Kenneth S., and Harold J. McMann. *U.S. Spacesuits*. Chichester, England: Springer/Praxis Publishing, 2006.

Young, Amanda J., and Mark Avino. *Spacesuits: The Smithsonian National Air and Space Museum Collection*. New York: Powerhouse Books, 2009.

Skylab

Compton, W. David, and Charles D. Benson. *Living and Working in Space: A History of Skylab*. Washington: NASA, 1983.

Cooper, Henry S. F., Jr. *A House in Space*. New York: Holt, Rinehart, and Winston, 1976.

Hitt, David, Owen Garriott, Joe Kerwin, and Alan Bean. *Homesteading Space: The Skylab Story*. Lincoln: University of Nebraska Press, 2008.

Launius, Roger D. *Space Stations: Base Camps to the Stars*. Old Saybrook, Conn.: Konecky & Konecky, 2003.

The Viking Lander

Boyce, Joseph M. *The Smithsonian Book of Mars*. Washington: Smithsonian Institution Press, 2003.

Digregorio, Barry E., Gilbert V. Levin, and Patricia Ann Straat. *Mars: The Living Planet*. London: Frog, Ltd., 1997.

Ezell, Edward Clinton, and Linda Neuman Ezell. *On Mars: Exploration of the Red Planet, 1958–1978*. Washington: NASA, 1984.

Launius, Roger D., ed. *Exploring the Solar System: The History and Science of Planetary Exploration*. New York: Palgrave Macmillan, 2013.

Sheehan, William. *The Planet Mars: A History of Observation & Discovery*. Tucson: University of Arizona Press, 1996.

Voyager

Dethloff, Henry C., and Ronald A. Schorn. *Voyager's Grand Tour: To the Outer Planets and Beyond*. Washington: Smithsonian Institution Press, 2003.

Evans, Ben, with David M. Harland. *NASA's Voyager Missions: Exploring the Outer Solar System and Beyond*. Chicester, England: Springer-Praxis, 2004.

Pyne, Stephen J. *Voyager: Seeking Newer Worlds in the Third Great Age of Discovery*. New York: Penguin, 2010.

Swift, David W. *Voyager Tales: Personal Views of the Grand Tour*. Reston, Va.: American Institute of Aeronautics and Astronautics, 1997.

Space Shuttle Discovery

Bizony, Piers. *The Space Shuttle: Celebrating Thirty Years of NASA's First Space Plane*. Minneapolis: Zenith Press, 2011.

Hale, Wayne, ed. *Wings in Orbit: Scientific and Engineering Legacies of the Space Shuttle, 1971–2010*. NASA SP-2010-3409. Washington: U.S. Government Printing Office, 2010.

Harland, David M. *The Space Shuttle: Roles, Missions and Accomplishments*. Chichester, England: John Wiley & Sons, 1998.

Jenkins, Dennis R. *Space Shuttle: The History of the National Space Transportation System*. Cape Canaveral, Fla.: Dennis R. Jenkins, 2001.

Jones, Thomas D. *Sky Walking: An Astronaut's Memoir*. New York: HarperCollins, 2006.

Hubble Space Telescope

DeVorkin, David, and Robert W. Smith. *Hubble: Imaging Space and Time*. Washington: National Geographic Books, 2008.

Kessler, Elizabeth A. *Picturing the Cosmos: Hubble Space Telescope Images and the Astronomical Sublime*. Minneapolis: University of Minnesota Press, 2012.

Smith, Robert W., with Paul A. Hanle, Robert H. Kargon, and Joseph N. Tatarewicz. *The Space Telescope: A Study of NASA, Science, Technology and Politics*. New York: Cambridge University Press, 1989. Paperback edition with Epilogue, 1993.

Weiler, Edward John. *Hubble: A Journey Through Space and Time*. New York: Abrams, 2010.

INDEX